MOLECULAR GENETIC ECOLOGY

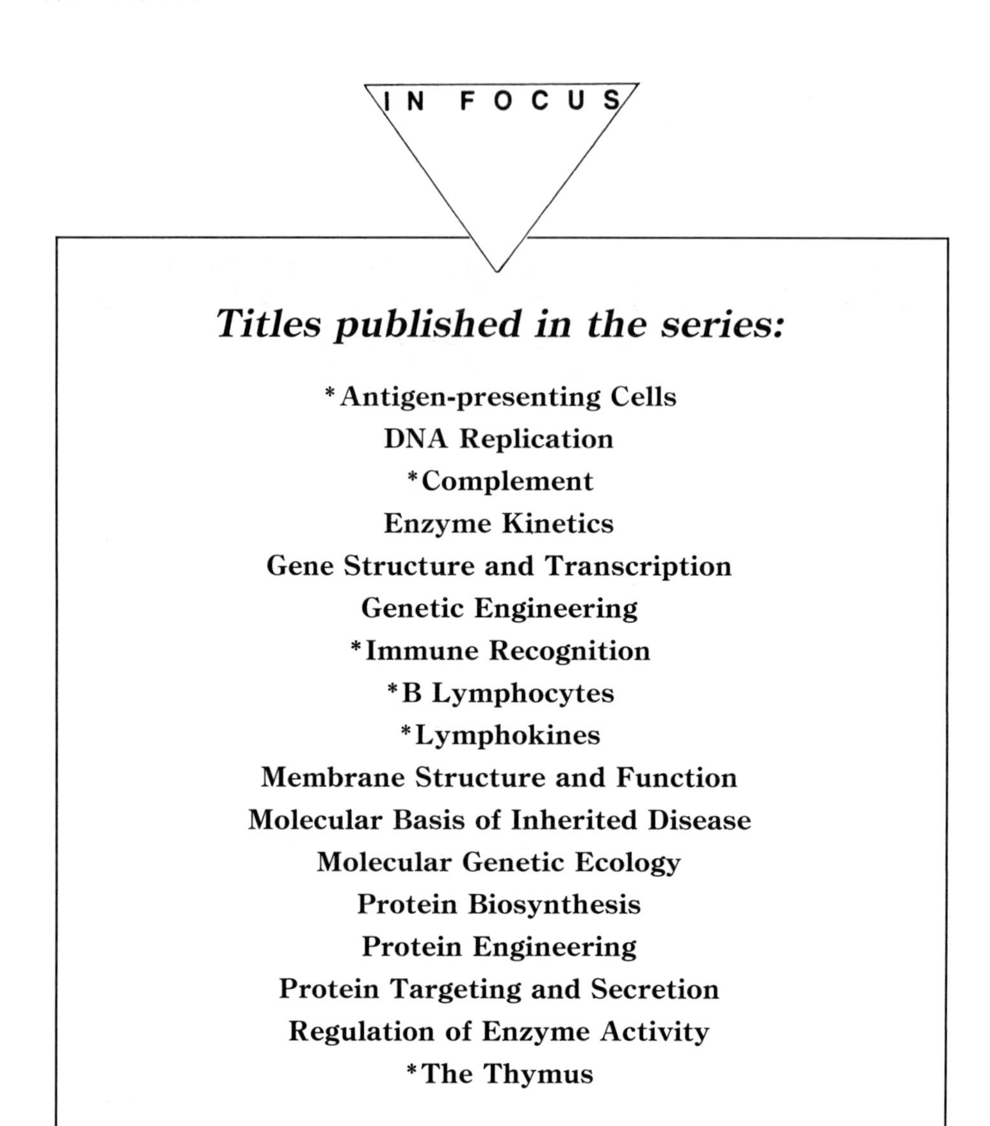

Titles published in the series:

***Antigen-presenting Cells**
DNA Replication
***Complement**
Enzyme Kinetics
Gene Structure and Transcription
Genetic Engineering
***Immune Recognition**
***B Lymphocytes**
***Lymphokines**
Membrane Structure and Function
Molecular Basis of Inherited Disease
Molecular Genetic Ecology
Protein Biosynthesis
Protein Engineering
Protein Targeting and Secretion
Regulation of Enzyme Activity
***The Thymus**

***Published in association with the British Society for Immunology.**

Series editors

David Rickwood

Department of Biology, University of Essex, Wivenhoe Park,
Colchester, Essex CO4 3SQ, UK

David Male

Institute of Psychiatry, De Crespigny Park, Denmark Hill,
London SE5 8AF, UK

MOLECULAR GENETIC ECOLOGY

A.Rus Hoelzel

Department of Genetics, University of Cambridge
Downing Street, Cambridge CB2 3EH, UK
and
Centre for Population Biology, Imperial College at Silwood Park, Ascot, Berkshire, UK

Gabriel A.Dover

Department of Genetics, University of Leicester, Leicester LE1 7RH, UK
formerly:
Department of Genetics, University of Cambridge
Downing Street, Cambridge CB2 3EH, UK

Oxford University Press
Walton Street, Oxford OX2 6DP

Oxford is a trade mark of Oxford University Press

Published in the United States
by Oxford University Press, New York

A catalogue record for this book is available from the British Library

Library of Congress Cataloging in Publication Data
Hoelzel, A. Rus.
Molecular Genetic Ecology / A.Rus Hoelzel, Gabriel A.Dover.
(In focus)
Includes bibliographical references and index.
ISBN 0-19-963265-0 (pbk.)
1. Population genetics. 2. Molecular genetics. 3. Variation (Biology). 4. Ecological genetics. I. Dover, G. A. (Gabriel A.) II. Title. III. Series: In focus (Oxford, England)
[DNLM: 1. Ecology. 2. Genetics, Biochemical. 3. Genetics, Population. 4. Molecular Biology. 5. Variation (Genetics)—genetics. QH 455 H694m]
QH455.H64 1991 575.1′5 – dc20 91-20807
ISBN 0-19-963265-0

Typeset and printed by Information Press Ltd, Oxford, England.

Preface

The advent of powerful molecular probes for uncovering naturally occurring genetic variation has greatly facilitated the understanding of behavioural ecology and the dynamics of populations in their ecological context. However, if the nature of a probe is not well understood, or if an inappropriate probe is used, then problems of interpretation can arise. It is important to couple a thorough understanding of the 'internal' mutational processes within nuclear and organelle genomes with an appreciation of the various 'external' processes of selection and drift that affect the standing level of genetic variation. In this book we briefly describe the underlying sources of genomic variation, how they can be detected and analysed, and how to interpret observed variation, taking into account the variety of contributing factors that affect its distribution.

A.Rus Hoelzel
Gabriel A.Dover

Contents

Abbreviations

APD	average percentage difference
HLA	human lymphocyte antigen
IGS	intergenic spacer
PCR	polymerase chain reaction
PGI	phosphoglucoisomerase
RFLP	restriction fragment length polymorphism
SDS	sodium dodecyl sulphate

1

Genetic variation in natural populations

1. Introduction

This chapter reviews the nature of mutations, and the ways in which they are disseminated through a population. Variation at the molecular level has been described and quantified only relatively recently, though it has long been recognized that individuals in a population vary phenotypically, and that some proportion of that variation is heritable. This book concentrates on the evolution of DNA, and the utility of measuring molecular variation for population level studies.

2. Natural populations

Species can be divided into groups of individuals that share kinship, as indicated by genetic or physical similarity. In some cases a large part of the diversity between individuals in a species is distributed in this way, between local 'populations'. Therefore, a genetic population can be usefully defined as a group of conspecific organisms that share greater kinship with each other than with the members of other similar groups. The forces that can genetically differentiate populations are genetic drift, natural selection, and DNA turnover mechanisms (see Section 5.3). Most often populations of this type are regional. That is, the isolating mechanism that has reduced the movement of breeding individuals (and therefore the transfer of genes) is usually distance, or some geographic barrier. The evolution of a population is largely dependent on its effective size (see Chapter 3; Section 6). This is the average number of individuals in a population that have equal genetic contributions to the subsequent generation. When population sizes fluctuate, or when non-random mating limits the number of reproductively active individuals in the population, the effective population size will be closer to the minimum number in the cycle, or the limiting number of reproducing individuals. This has the effect of reducing the level of genetic variation in the population.

3. Nature of mutations

Heritable genetic variation accumulates by mutation at the DNA level. The DNA molecule, which consists of a long chain of paired nucleic acid bases, can be altered in a variety of ways. The best understood involves the substitution of one base for another (point mutation). Point mutations accumulate primarily by copy error (during replication) and by induced mutation (from 'mutagens' in the environment).

Other types of mutation involve the deletion, addition, or replacement of stretches of DNA (which may vary in length from one to several thousand nucleotides); the inversion of a length of DNA at a given position; and the movement of lengths of DNA from one position to another. Such mutations are caused by mechanisms of DNA turnover known as slippage, unequal crossing over, gene conversion, and transposition, amongst others. Their modes of operation are described briefly below.

Variation in a population also arises as a consequence of the sexual process. This is due to the independent assortment of non-homologous chromosomes, and crossing-over between homologous chromosomes.

4. Mutation rates

Eukaryotic genes in nuclear genomes are split into coding segments (exons) separated by stretches of non-coding DNA (introns) which are transcribed into RNA, but are not translated into a polypeptide chain (*Figure 1.1*). Further, genes begin and end with transcribed but non-translated flanking sequences. Li and co-workers (1) have compared substitution rates between these various regions

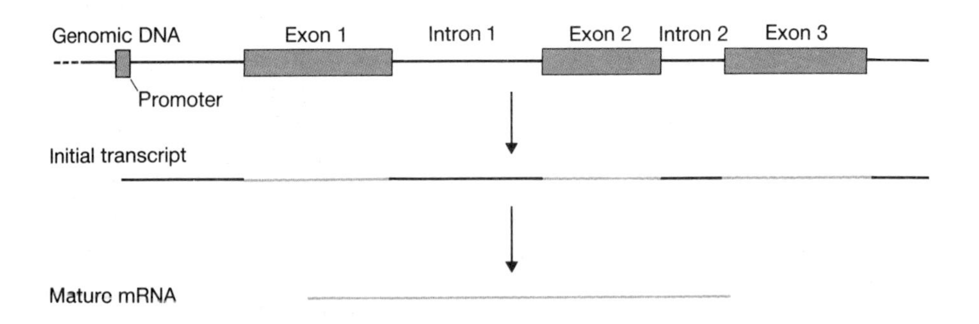

Figure 1.1. Division of nuclear genes into coding (exon) and non-coding (intron) regions. The initial transcript across all regions is modified to represent only the exon regions in the mature messenger RNA.

and found that intron and flanking regions evolve at a considerably higher rate than do coding regions.

In general, the lowest rates of nucleotide substitution occur in the exons of genes. However, within this category rates are highly variable. They range from 0.004×10^{-9} substitutions per year in the gene for histone H2A to 2.8×10^{-9} in the gene for interferon A (see 2 for a review on the range of non-synonymous substitution rates in the first and second positions of codons). The rate for histone H2A is equivalent to only one or two amino acid changes between a pea and a cow, while interferon A would be so different between these two taxa that the sequence would be hardly recognizable. The average rate for mammals is 0.88×10^{-9} substitutions per non-synonymous site per year. The substitution rate at synonymous sites has been suggested to vary from 1.7 to 11.8×10^{-9} substitutions per year. The average is about 5×10^{-9} substitutions per year, which is five times higher than average non-synonymous substitution rates. The effect of inconsistent substitution rates on models for quantifying population diversity are discussed below.

Many genes in eukaryotic nuclear genomes exist in multiple copies (multigene families) or are internally repetitive, due to repeated duplications by unequal crossing-over and similar amplification mechanisms (see Section 5.3.2). The rate of divergence in multigene families (such as those coding for histones, immunoglobulins, and ribosomal RNAs) is complicated by the fact that a mutation occurring in one member gene can spread to other copies of the family, by any one of several DNA turnover mechanisms. The family, or some subsection of it, can become fully or partially homogenized, depending on the rate of spreading compared to the rate of mutation. If new mutations spread through the gene family faster than they are acquired, then the gene family will evolve as a unit (3). Multiple genes that are in a tandem array are often separated by spacers which experience weaker functional constraints than the genes themselves. Ohta (4) has compared rates of divergence within the immunoglobulin multigene family between humans and rabbits. She describes a rate of 0.7×10^{-9} substitutions per nucleotide site per year in the coding regions and 1.8×10^{-9} for the spacer regions (which corresponds to the high end of the spectrum for structural genes).

The highest known evolutionary rates are in repetitive DNA sequences such as the hypervariable minisatellite regions described by Jeffreys and co-workers (5). Nucleotide substitution rates in these regions are over an order of magnitude higher than the average for coding regions (about 2.0×10^{-7}). Rates of change in the length of tandemly repeated families (like minisatellites) can be considerably higher, leading to individual specific lengths ('genetic fingerprints'). This is due to DNA turnover mechanisms, as described in Section 5.3.

5. Dissemination of mutations in populations

Essentially, there are three ways by which mutations can spread in a sexual population. These are genetic drift, natural selection, and molecular drive.

Genetic drift is a change in gene frequencies that is a consequence of the continual random gain and loss of gametes and individuals in a population. Although genetic drift occurs in populations of all sizes, the effect is most significant in very small populations. Natural selection is a consequence of differences in the extent to which genetically distinct individuals interact with their environment: an interaction affecting their relative reproductive success. Molecular drive (3) is a term used to describe the spreading consequences of various mechanisms of DNA turnover that are operationally distinct from selection and drift.

Selection, drift, and molecular drive are all independent but superimposed one upon another, leading to complex patterns of population change and differentiation. Unravelling all three processes becomes a necessity in trying to assess the nature and significance of genetic variation, in particular at the DNA level. Fortunately, the difficulty of quantifying the relative contributions of the three processes to observed diversity in any given genomic component does not always hamper the exploitation of this diversity for the correct identification of hierarchical levels of taxa.

5.1 The neutral theory

The neutral gene hypothesis (6) suggests that mutation is the primary force in evolution (see 7). According to this theory, evolution occurs by the random fixation of neutral or nearly neutral mutations. At any one time, the degree of polymorphism is a consequence of new variation tending towards fixation or elimination by chance. Deleterious alleles are removed by selection. The neutral theory predicts high levels of protein polymorphism in natural populations, and

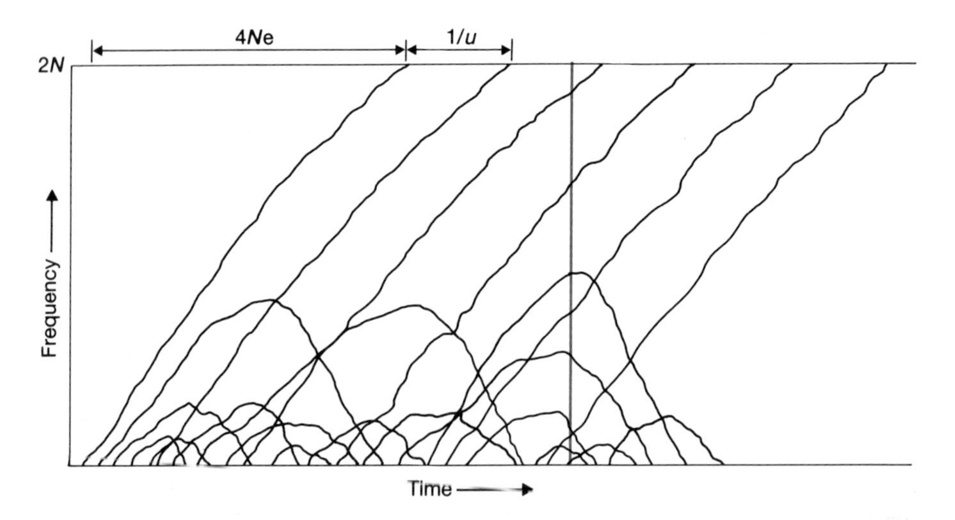

Figure 1.2. Neutral evolution. The neutral theory predicts that a mutation will go to fixation every $1/u$ years and require $4N_e$ generations. Many mutations will die out before reaching fixation. At a given point in time, there will be many genes on their way to fixation or loss, and therefore polymorphic in the population (represented by intersections with the orange line in the figure).

approximate constancy in the rate of amino acid substitutions for each protein. Once a new allele enters the gene pool it is either lost by chance or goes to fixation (at a probability of $1/2N_e$, where N_e is the effective population size). The number of new alleles that are fixed each generation is the same as the mutation rate, u, and the time it takes them to go to fixation is approximately $4N_e$ generations. Therefore, at any point in time there should be numerous alleles 'on their way' to fixation or loss, not yet represented in all individuals. This is why the neutral theory predicts high levels of polymorphism (*Figure 1.2*).

Variation can be reduced in natural populations by two stochastic mechanisms, consistent with the neutral hypothesis: the founder effect (a single gravid female colonizing a new area; 8) and the 'bottleneck' phenomenon (9). In either case the population is reduced to a small inbreeding group where loci are readily fixed by genetic drift. From this condition of reduced heterozygosity, the process of gradual accumulation of new mutations would recommence. This process is very slow to reach an equilibrium level, approximately the reciprocal of the mutation rate (potentially longer than the life of a species) (9).

5.2 *Natural selection*

An alternative view to the neutral theory is that natural selection is the dominant creative force in evolution. Variation in natural populations is thought to be maintained by balancing selection and overdominance (heterozygote advantage) (see 10). Directional selection can limit variation by increasing the representation of one phenotype at the expense of another, or increase variation in a heterogeneous environment if one allele is most appropriate in a local region, and another in a different region. Overdominance can increase variation by conferring an advantage to individuals that have two different alleles at a locus, and thereby maintain both alleles in the population.

Adaptation to specific habitats can affect levels of diversity in general. The niche-variation model of selection (11) suggests that specialized organisms occupying a narrowly-defined niche will exhibit low genetic variability. Another theory contrasts 'fine-grained' and 'coarse-grained' environments (12). Fine-grained environments vary seasonally, but are predictable over time and consistent for all individuals. Coarse-grained environments vary randomly from one generation to the next. On the basis of this model, mobile animals with the capacity to select favourable environments and the physiological mechanisms to adapt to variations in the environment (e.g. mammals) will have low levels of heterozygosity in fine-grained environments. Smith and Fujio (13) surveyed 106 teleost fish species and found that generalist species adapted to fine-grained environments had lower levels of variation than habitat specialists.

It has been hypothesized that the low levels of heterozygosity found in some fossorial mammals, the American alligator, *Alligator mississippiensis*, and the harp seal, *Pagophilus groenlandicus*, reflect the relative stability of their environments. Tolliver *et al.* (14) compared the genetic diversity in fossorial and terrestrial species within the class Insectivora. They found that fossorial species were less variable, but that terrestrial insectivores were less variable than fossorial rodents.

However, the findings of various researchers suggest that exceptions are common (e.g. benthic marine invertebrates exhibit relatively high levels of genetic variation; 15) and that a combination of genetic factors will better predict heterozygosity.

5.3 DNA turnover mechanisms and molecular drive

A decade of investigations into the organization of eukaryotic nuclear genomes has revealed a variety of molecular mechanisms of DNA turnover operating in all examined species embracing the major living kingdoms. Such mechanisms can produce new types of mutation and can be involved with the dissemination of the mutations (molecular drive) through sexual populations (see examples given below in this section). All mechanisms cause the gain or loss of genetic variants in the lifetime of an individual. Such small but persistent patterns of non-Mendelian segregation can affect the genetic composition of a population over long periods of time (3). For example, if a turnover mechanism generates a new variant copy of a gene on a chromosome different from the original gene, then the sexual process of meiosis and gametic fusion can ensure that the original and variant genes enter different individuals at the next generation. The continual repetition of gain and loss of genes in the lifetime of an individual, when coupled to sex, will ensure that a novel gene may either spread and replace all copies of the original gene, or disappear from the population. This process is analogous to the diffusion of alleles through a population by neutral drift, except in cases where a turnover mechanism can be biased in favour of a new variant gene, as in biased gene conversion or transposition (see next section).

It is important to discriminate between the means turnover mechanisms provide for spreading variation, and their ability to generate variation (see Section 5.3.4 and Chapter 4, Section 4). In general, turnover mechanisms are responsible for generating variation at a significantly higher rate than variation due to point mutations. Consequently, methods that investigate this type of variation offer considerable potential for kinship studies (see Chapter 4, Section 6).

5.3.1 Transposition

This is the best understood mechanism. It involves the movement of a length of DNA from one position to another in the genome. The lengths can vary, depending on the mobile element involved, from approximately 0.5 to 8 kilobases (kb). There are two types of transposition: the first involves the duplication of a length of DNA followed by its insertion elsewhere (duplicative transposition); the second simply involves excision and reinsertion (non-duplicative transposition). Variation induced by these mobile elements can take several forms. If the element moves into a gene or its flanking controlling sequences then a mutation could ensue. Further, it is known that excision of an element can be imprecise in that a few extra bases may be left behind as a 'footprint' at the site of excision. Such 'footprints' can affect the expression of a gene and lead to the gradual generation of multiple alleles at a locus (16). If an element moves into non-coding sequences (of which there are many in most genomes), then the length of that region will be altered. This is detectable by standard molecular methods (see Chapter 2). In general, the rates of transposition are low (from 10^{-2} to 10^{-4} events per generation). Hence, different populations can be identified by different but relatively invariant positions of a family of mobile elements.

5.3.2. Unequal crossing-over

Unequal crossing-over can occur between two chromatids or between two chromosomes (homologous and sometimes non-homologous), when there is incomplete alignment between the two structures. After crossing-over, one structure gains extra genetic material and the other suffers a corresponding loss (*Figure 1.3a*). Unequal crossing-over is a means of generating duplicate genes,

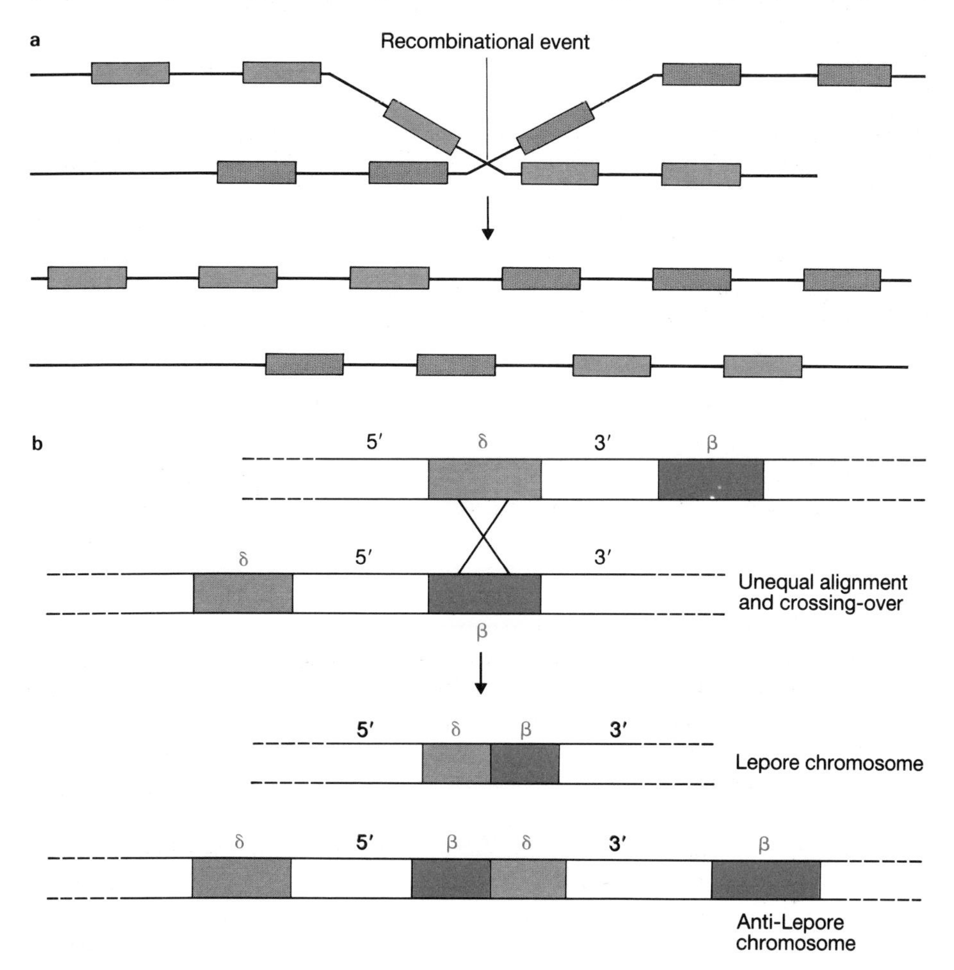

Figure 1.3. Unequal crossing-over. **a** The general case for the generation or loss of repeats in a tandem array. **b** An example of unequal crossing over in the β-globin cluster of genes in humans generating short (Lepore) and long (anti-Lepore) recombined chromosomes. Only the δ and β genes are shown.

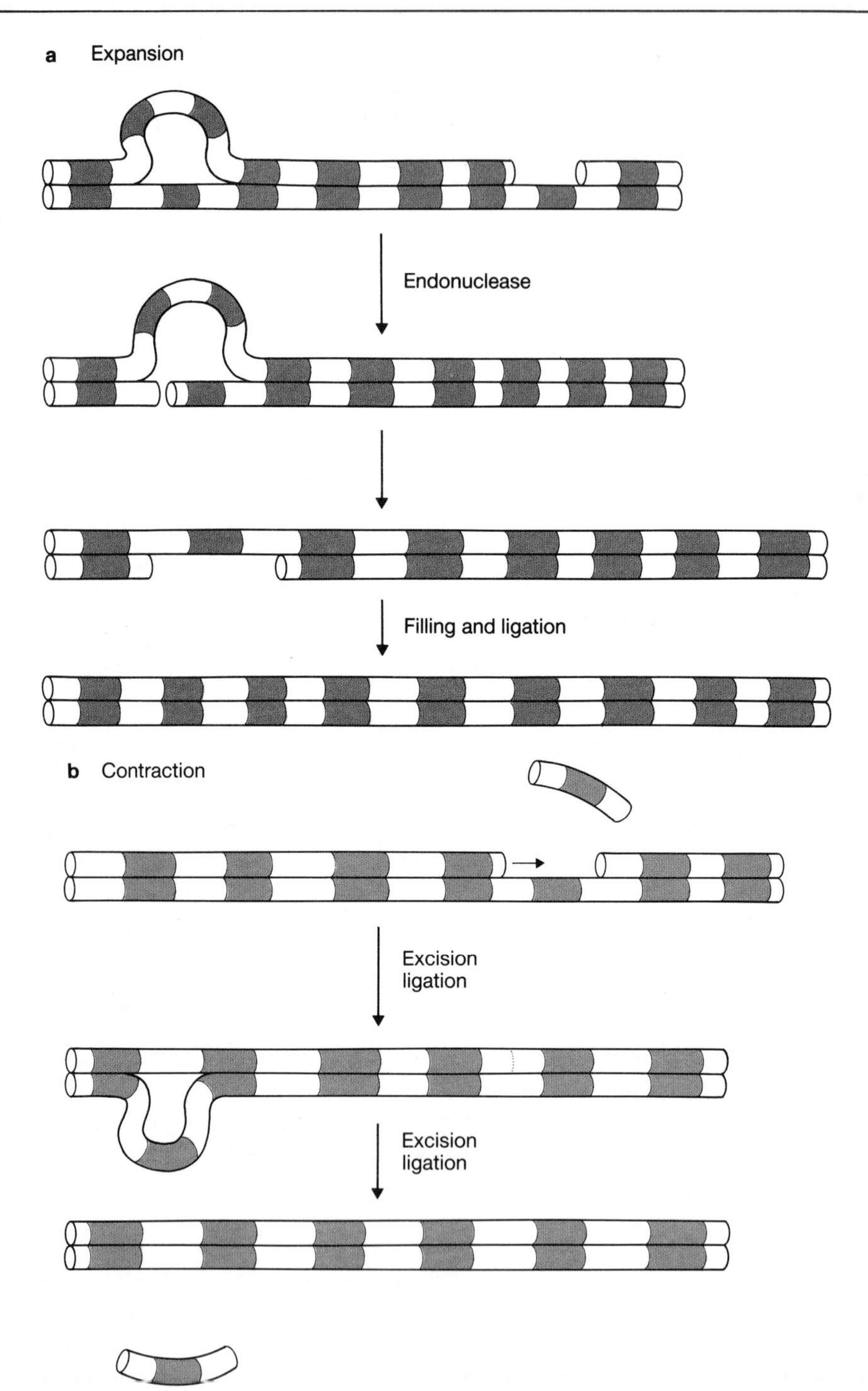

Figure 1.4. DNA slippage as a source of variation in repeat numbers by DNA repair: **a** expansion **b** contraction.

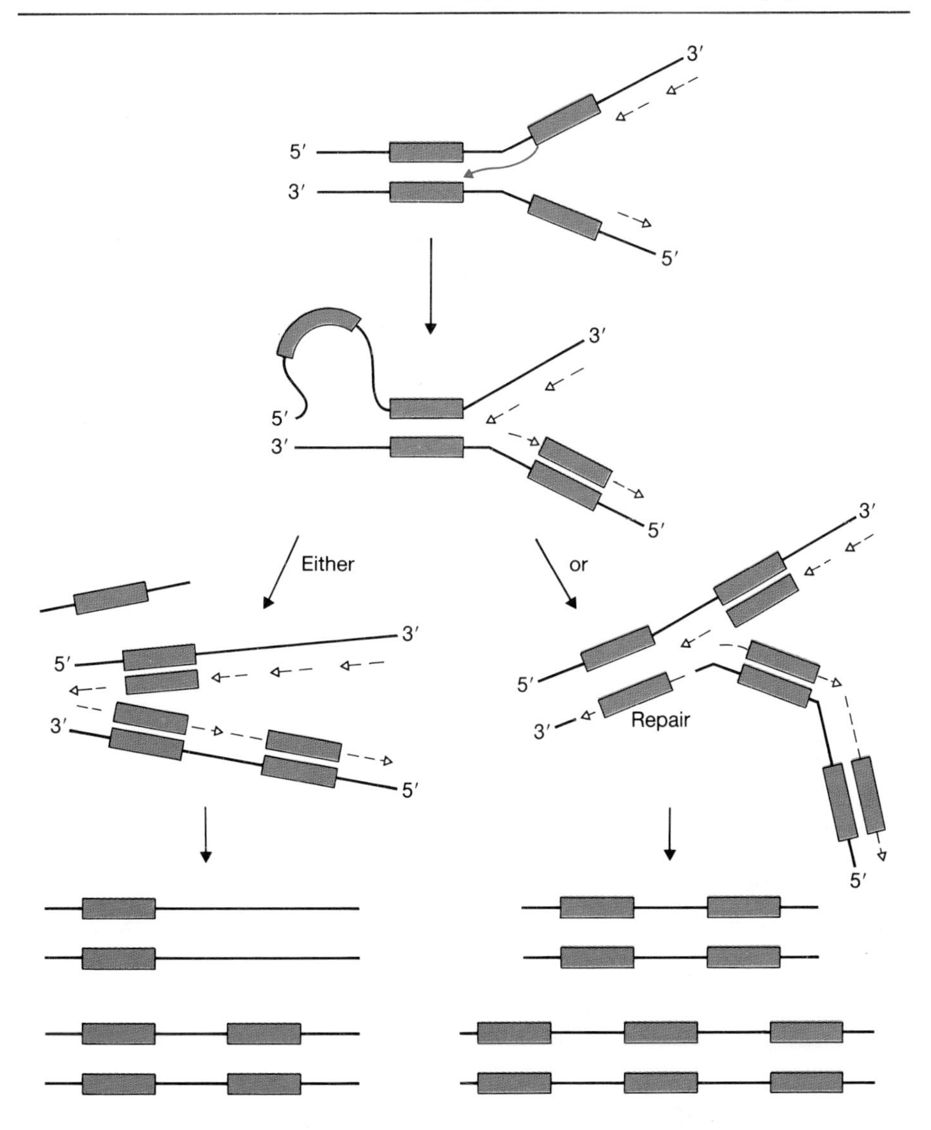

Figure 1.5. Alteration in the number of tandem repeats by slippage replication. In this model, either the looped-out portion breaks off resulting in a contraction of copy number, or the opposing strand breaks followed by DNA polymerase repair resulting in the expansion of copy number.

as seen, for example, in some of the mammalian globin genes (*Figure 1.3b*). The continual activity of unequal crossing-over can produce further rounds of duplication, leading to long tandem arrays of a given gene and hence generating a multigene family. There are many gene families that have been generated in this way, the most notable being the genes for the ribosomal RNAs (e.g. 18S, 28S, and 5S RNAs), the five histone genes (often in repetitive units each of which

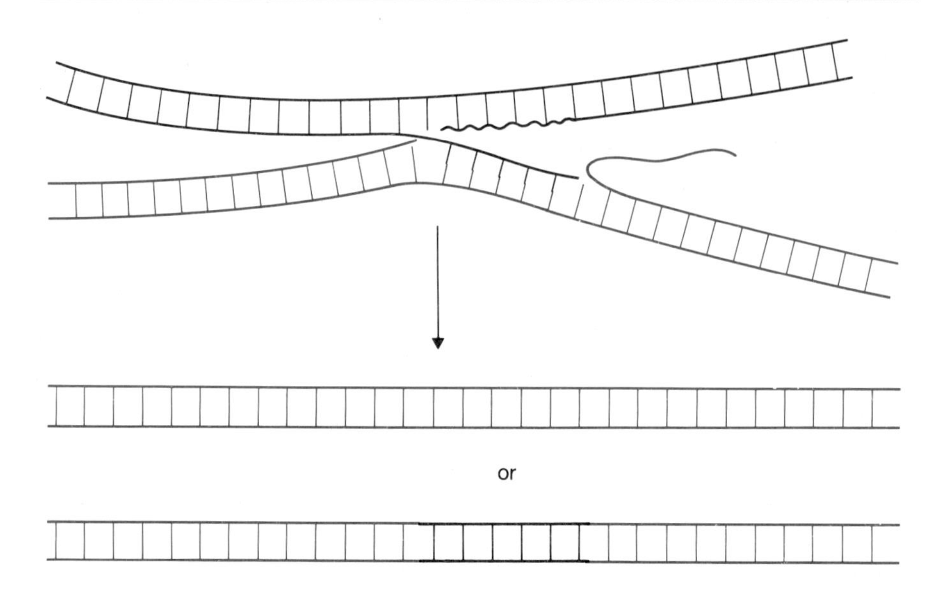

Figure 1.6. Gene conversion. The DNA is repaired in favour of either the infiltrating or the original sequence.

contains all five genes), and the several 'variable' and 'constant' genes of the mammalian immune superfamily of genes (see 17).

Once a gene family is established, unequal crossing-over is involved with its maintenance, in the sense that the mechanism ensures that genetic variation between member genes is continually reduced (4). For example, if a mutation occurs in one of a hundred genes in a pair of arrays of an individual, then there is some probability that after many rounds of unequal crossing-over, the mutant gene will replace all the original arrays in the population.

5.3.3. Slippage

The precise molecular events of slippage (sometimes known as slippage-replication) are not known. It is recognized, however, that many regions of nuclear genomes are composed of very short (on average less than 10 bp) motifs of DNA that are in tandem arrays (pure simplicity) or scrambled one with another (cryptic simplicity) (see 18). The numbers of copies of any given motif in any defined region are much higher than would be expected to occur by chance in a random sequence of the same length and composition of nucleotides. Both pure and cryptic simplicity are considered to be due to the propensity for the two strands of the DNA helix to slip against each other, creating a gap on one side and a buckle (loop) on the other. Repair of such lesions or structures can lead to a gain or loss of short motifs of DNA (*Figures 1.4* and *1.5*). Detailed computer analyses of many genes, when compared between species, indicate that slippage-generated variation is widespread and that it is being produced at a faster rate than the rate of point mutations.

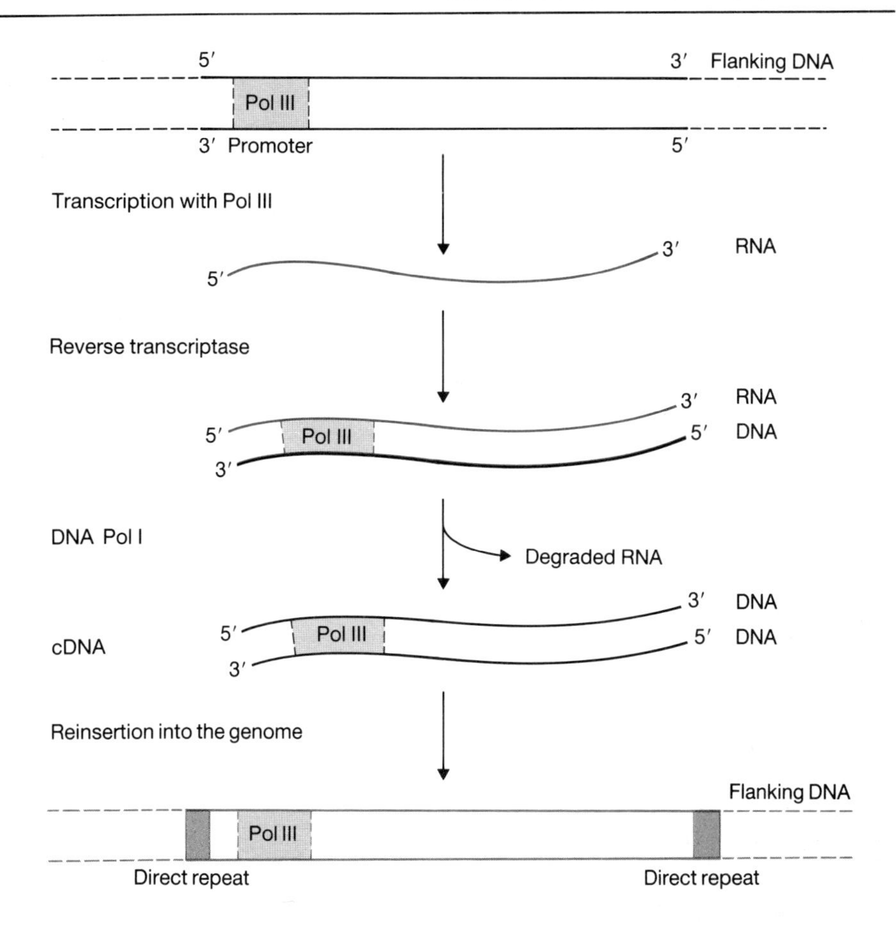

Figure 1.7. RNA-mediated turnover via reverse transcription. Pol III transcribed genes have a promoter inside the gene (example shown). In this case, the promoter is transposed with the rest of the sequence, allowing the process to continue. Genes transcribed by Pol II have the promoter outside the gene on the 5′ side (example not shown), which prevents repeated transposition of the same sequence. When the transposable element inserts into a new site, direct repeats are generated at either side of the point of insertion.

5.3.4 Gene conversion

Analysis of non-Mendelian patterns of gene segregation in meiotic tetrads of fungal species has revealed the phenomenon of gene conversion (19). This is a mechanism which involves the non-reciprocal transfer of sequence between copies (alleles or non-alleles) of a gene. That is, starting with two slightly different copies, gene conversion leads to two identical copies. This process is thought to be due to the invasion of a double helix of one member gene by a single strand of the helix of another. After much twisting and turning, the resultant heteroduplex is repaired to give rise to a stably base-paired homoduplex (*Figure 1.6*). The direction of repair can be arbitrary (unbiased gene conversion) in which case a heteroduplex of composition *Aa* can be repaired to either *AA* or *aa*. If the repair is more frequently in the direction of either *A* or *a* then it is said to be biased.

Gene conversion can involve regions of DNA from just a few bases to tens of thousands of bases. Although by definition it is a process of homogenization, nevertheless it can lead to genetic variation if the unit of DNA under comparison between taxa is longer than the gene conversion domains within it. In such an instance the unit of DNA becomes a mosaic of different conversion domains, and each unit in the separate taxa is a differently composed mosaic. Much of the high variability in the several genes involved with the mammalian immune system has arisen by such a disparity between the length of the gene and the conversion domains (see 20).

5.3.5 RNA-mediated transfers of genetic information

Some proportion of the available genetic variation in nuclear genomes is due to the turnover of DNA sequences via their RNA intermediates. This is a consequence of the presence of reverse transcriptase which transcribes RNA into its complementary DNA (cDNA), followed by the reinsertion of the cDNA into the genome at many different loci. Many processed pseudogenes (i.e. genes without introns and the 5′ and 3′ control sequences) arise in this way. For example, there are several copies of the α-tubulin gene in the rat nuclear genome, but only one is functional, the others are processed pseudogenes. Some very large DNA families such as the 500 000 copies per individual human of the 'Alu' family, in addition to many other repetitive families, arise via their RNA intermediates (*Figure 1.7*). Such a mechanism can lead to a relatively rapid accumulation of repetitive elements and pseudogenes because of the vast numbers of RNAs per nucleus, leading to large differences in the copy-number of a given repetitive family, even between closely-related species.

6. Further reading

The neutral theory

Kimura,M. (1983) *The Neutral Theory of Molecular Evolution.* Cambridge University Press.

Natural selection

Lewontin,R.C. (1974) *The Genetic Basis of Evolutionary Change.* Columbia University Press, New York.

Molecular drive

Dover,G.A. and Flavell,R.B. (ed.) (1982) *Genome Evolution.* Academic Press, London.

7. References

1. Li,W.-H., Wu,C.-I. and Lou,C.-C. (1985) *Mol. Biol. Evol.*, **2**, 150.
2. Li,W.-H., Lou,C.-C., and Wu,C.-I. (1985) In *Molecular Evolutionary Genetics.* MacIntyre,R.M. (ed.), Plenum Press, New York, p.1.

3. Dover,G.A. (1982) *Nature*, **299**, 111.
4. Ohta,T. (1980) *Evolution of Multigene Families*. Lecture Notes on Biomathematics. Springer-Verlag, Berlin.
5. Jeffreys,A.J., Wilson,V., and Thein,S.L. (1985) *Nature*, **314**, 67.
6. Kimura,M. (1968) *Genet. Res.*, **11**, 247.
7. Nei,M. (1983) In *Evolution of Genes and Proteins*. Nei,M. and Koehn,R.K. (ed.), Sinauer, Sunderland, MA, p.165.
8. Mayr,E. (1963) *Animal Species and Evolution*. Harvard University Press, Cambridge, MA.
9. Nei,M., Maruyama,T., and Chakraborty,R. (1975) *Evolution*, **29**, 1.
10. Hartl,D.L. (1980) *Principles of Population Genetics*. Sinauer, Sunderland, MA.
11. Levine,H. (1953) *Amer. Nat.*, **87**, 331.
12. Levins,R. (1968) *Evolution in Changing Environments*. Princeton University Press, Princeton, NJ.
13. Smith,P.J. and Fujio,Y. (1982) *Marine Biol.*, **69**, 7.
14. Tolliver,D.K., Smith,M.H., and Leftwich,R.H. (1985) *J. Mammal.*, **66**, 405.
15. Ayala,F.J., Valentine,J.W., Hedgecock,D., and Barr,L.G. (1975) *Evolution*, **29**, 203.
16. Coen,E.S., Carpenter,R., and Martin,C. (1986) *Cell*, **47**, 285.
17. Dover,G.A. (1986) *Trends in Genetics*, **2**, 159.
18. Tautz,D., Trick,M., and Dover,G.A. (1986) *Nature*, **322**, 652.
19. Whitehouse,H.L.K. (1983) *Genetic Recombination—Understanding the Mechanism*. Wiley, New York.
20. Dover,G.A. and Strachan,T. (1987) In *Evolution and Vertebrate Immunity*. Kelsoe,G. and Schulze,D.H. (ed.), University of Texas Press, Austin, TX.

2

Molecular approaches to the analysis of genetic variation

1. Introduction

The existence of DNA sequence variation can be a powerful tool for the characterization of populations. A number of techniques have been developed to assess levels of genetic variation. The degree and type of variation depends very much on what part of the genome is being investigated. Examining variation in proteins (a reflection of changes in the genes that code for them) has been by far the most common technique. The strength of this approach is its emphasis on DNA sequences that are expressed phenotypically. This allows an investigation into the role of selection in the evolution of gene loci. The procedure is also inexpensive and relatively easy to conduct (especially horizontal starch gel electrophoresis). However, protein studies investigate only some of the variation in the most conserved class of DNA (usually single-copy coding sequences), and this limits the amount of possible resolution of time-dependent variation. The highly variable segments within the structural gene (introns, flanking sequences, and synonymous third-position codon sites) are not detected by this technique.

A number of new recombinant DNA techniques are now available for the analysis of more variable regions of the genome. Often these procedures involve the isolation of a DNA fragment which is then radioactively labelled and used to 'probe' the genome for similar sequences. Genomic DNA is 'cut up' with restriction enzymes and assorted by size in an electric field on a horizontal gel. The DNA in the gel is then transferred to a filter which is probed with the labelled fragment. These and other procedures are described in more detail in the following sections.

2. Enzyme electrophoresis

By measuring the migration of proteins through an electric field in a gel matrix (gel electrophoresis) different alleles of the same gene can often be distinguished.

Proteins are charged molecules that will migrate in an electric field, and proteins of different composition and conformational properties will migrate at different rates. Enzymes are detected by immersing the gel into a medium containing the reaction conditions for the enzyme such that it incorporates a coloured dye into the end product.

The apparatus for gel electrophoresis can be easily built from Perspex (plexiglass) and two platinum wire leads. Gels can be prepared from a variety of media (starch, polyacrylamide, agar, etc.) and run by various methods (horizontal, vertical, disc, etc.). For screening large numbers of loci or samples in a population study, horizontal starch gels are usually preferred (*Figure 2.1*). This is because starch is inexpensive and non-toxic. Gels running 10–30 lanes can be sliced horizontally four times and then each slice stained for a different enzyme. Resolution for enzymes is often as good as with other media. Poor resolution can sometimes be improved with polyacrylamide gels or isoelectric focusing (see 1 and next paragraph). Procedures for running and staining gels are outlined in detail elsewhere (2–4).

Characteristics of the running buffer and supporting medium both affect resolution. This is often a result of the effect different media and combinations of buffer have on pore size. Discontinuous buffer systems are used in polyacrylamide gels to create a stacking effect at the boundary between regions of large and small pore size, which improves the resolution of the bands in the second (small pore) phase. In isoelectric focusing, proteins migrate through a pH gradient to the point where they become electrically neutral, referred to as the isoelectric point (pI) of the protein. If the protein diffuses from this point it will develop a charge and return to its pI point, a process which has the effect of concentrating or 'focusing' the protein into a very narrow band. Further resolution can be obtained by running a gel in two dimensions (running it once, and then again at a right angle to the first run). A common combination is isoelectric focusing in the first dimension, followed by SDS electrophoresis in the other dimension. Polypeptides run through gels containing SDS (sodium dodecyl sulphate) separate mainly according to their molecular mass.

A number of factors can influence the interpretation of allozyme polymorphisms when enzymes are stained on the basis of their activity. Many enzymes are composed of more than one polypeptide. These 'multimeric' enzymes show

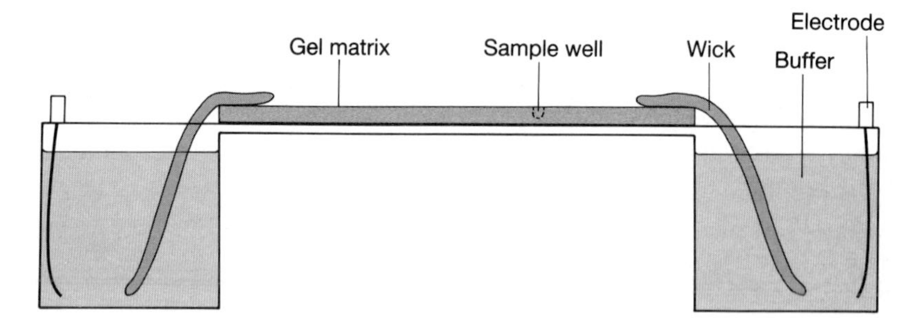

Figure 2.1. Non-denaturing gel electrophoresis of proteins.

hybrid bands in the heterozygous condition. When staining intensities are symmetrical between all alleles in the heterozygote, the pattern is easily interpreted. However, sometimes alternative alleles have different enzymatic activities (and therefore staining intensities). In the extreme case, when an allele has no enzymatic activity, the heterozygote will appear identical to the homozygous genotype for the active allele. The so-called 'null' allele can be detected when the genotype for the inactive allele is homozygous. Also, when a null allele is present there will be an apparent deficiency of heterozygotes in the population compared to Hardy–Weinberg expectations (see Chapter 3).

Sometimes more than one locus codes for the same protein. In many cases the gene products from the different loci can be easily distinguished, but confusion can arise if the products of multiple loci overlay one another on the gel. It is sometimes possible to separate the loci by using different substrates for activity staining, changing the pH of the gel buffer, or selectively inhibiting the activity of some loci (3,4). Some enzymes which are relatively non-specific, such as esterases, are especially prone to this problem.

Another potential problem comes with the interpretation of bands that result from post-translational changes in the structure of the protein. Harris and Hopkinson (4) discuss the potential causes of these 'secondary' isozyme patterns. The changes are often by oxidation of sulphydryls, deaminations, or acetylations and can occur either *in vivo* or *in vitro* during storage, extraction, or electrophoresis. Proper storage and handling can help minimize the artifactual formation of secondary isozyme effects.

To establish that allelic patterns are consistent with Mendelian inheritance, it is best to conduct breeding experiments. As this is clearly impractical with some species, consistency with known biochemical properties of the protein and agreement with Hardy–Weinberg expectations (see Chapter 3, Section 1) should be established as indirect lines of evidence. Having supported the allelic basis of the allozyme polymorphisms, it is important to establish whether alleles at different loci assort independently. Independence can be determined by comparing allelic distributions for all loci. Many statistical analyses depend on independence between all pairs of loci. In the absence of selection or pleiotropy (one gene having multiple effects), non-random assortment could indicate linkage (proximity of the loci on the same chromosome) or some form of non-random mating, such as polygyny or inbreeding.

3. Nucleic acid electrophoresis, restriction enzymes, and Southern blotting

Whole-cell DNA is extracted from tissue by lysing the cells, separating proteins and carbohydrates from nucleic acids, and precipitating the DNA with salt and 100 per cent ethanol (*Figure 2.2*). First, cellular material (for example, blood or tissue from animals, leaves, or roots from plants) is homogenized in a lysis buffer with a protease, an enzyme that breaks down proteins. Then an equal volume of phenol is added and the mixture shaken fairly vigorously. Phenol and water

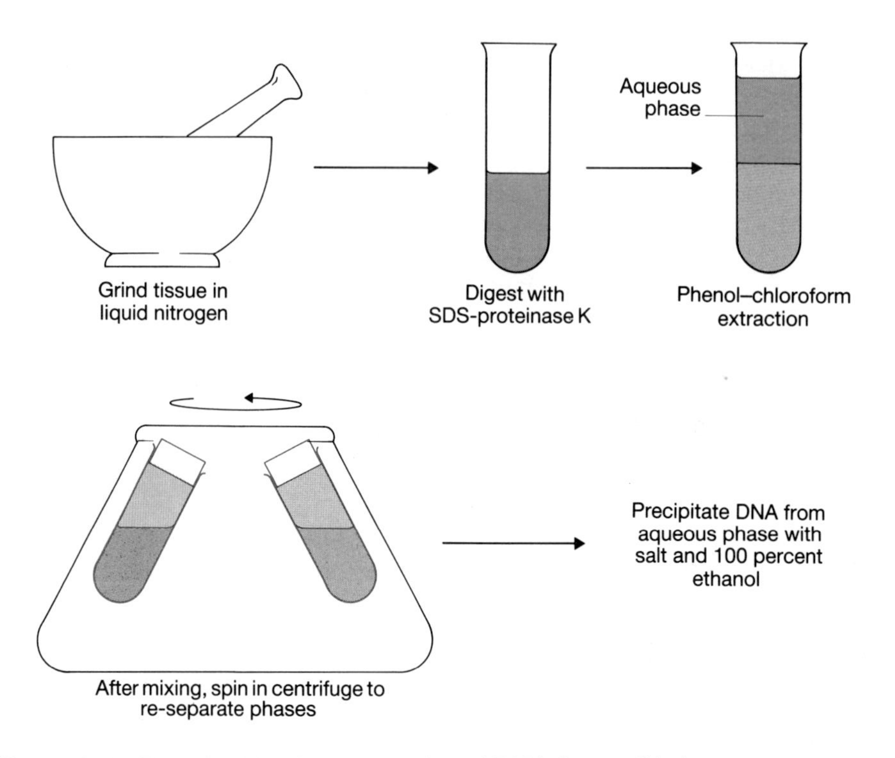

Figure 2.2. Phenol–chloroform extraction of DNA from solid tissue.

are immiscible. Due to their respective charge properties, proteins and amino acids migrate into the phenol phase and interface, while nucleic acids remain in the aqueous phase. This process is repeated, and then excess phenol is removed by extracting the aqueous phase with chloroform. DNA is precipitated from the aqueous phase with salt and ethanol.

A preparation of DNA that has been carefully extracted will be composed predominantly of very large fragments: 20–100 kb long. Excessive agitation can shear DNA into smaller fragments. The DNA can then be cut in a predictable and reproducible way with a class of enzymes known as restriction endonucleases. These enzymes recognize specific sequences, usually four, five, or six base pairs long, and cleave the DNA at every incidence of that sequence (*Figure 2.3*). They are prokaryotic enzymes which serve, for example, to defend a bacterium against invading viral DNA. A bacterium may protect its own DNA by methylating the restriction sites, so that the endonuclease no longer recognizes them. Some endonucleases are sensitive to methylation and some are not. For example, *Hpa*II and *Msp*I are 'isoschizomers' (they both recognize the same sequence of DNA: CCGG), but *Hpa*II will not cut at methylated sites, while *Msp*I is insensitive to methylation.

In a random DNA sequence, a restriction enzyme that recognizes a 4-bp site will cut every 256 bp on average. An enzyme with a 6-bp recognition site will

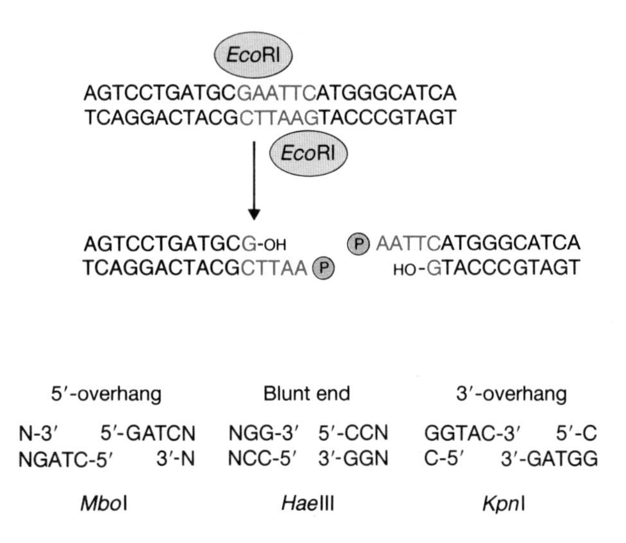

Figure 2.3. Restriction enzyme digestion of DNA.

cut every 4096 bp. This is a consequence of the fact that a particular sequence of DNA will occur at a certain frequency in a random stretch of DNA just by chance (see Section 5). These enzymes are highly specific and predictable, and most are commercially available.

DNA is usually electrophoresed through a dilute gel of agarose. Horizontal agarose gels are formed in a rectangular plate and run under a buffered solution in an electric potential difference of 20-150 V (*Figure 2.4*). To make the gel, molten agarose is poured into a casting plate around the teeth of a 'comb' set at one end of the plate. When the gel has set, the comb is removed leaving a series of rectangular troughs or 'wells' into which the samples are pipetted. A marker dye containing sucrose or Ficoll is added to the samples to cause them to sink into the wells.

DNA migrates through the gel in the anodal direction (from − to +) at different rates according to the size of the fragment. The relationship is roughly exponential, with large fragments moving through the gel disproportionately slowly, but this depends to some extent on the concentration of agarose in the gel. Low-concentration gels resolve large fragments best. The DNA can be visualized under UV irradiation after staining with ethidium bromide, a molecule which intercalates into nucleic acids and fluoresces bright orange in UV light.

Eukaryotic genomic DNA cut with a restriction enzyme will appear as a smear on a gel stained with ethidium bromide, because there will be so many different-sized fragments. In some cases, there will be bands representing repetitive regions of DNA that comprise a significant proportion of the genome (in copies of up to millions of repeats). For phylogenetic comparisons it is useful to be able to investigate the chance loss and gain of restriction sites in a particular gene or region of DNA. These are referred to as restriction fragment length

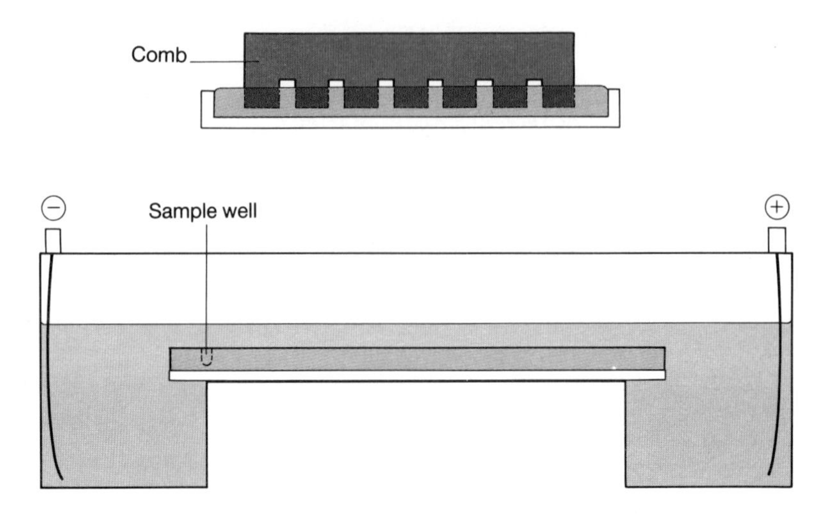

Figure 2.4. Submarine agarose gel electrophoresis of DNA restriction fragments.

polymorphisms (RFLPs). In order to do this it is necessary to distinguish the region of interest from the background smear of genomic DNA. The first step is to transfer the DNA in the gel to a nitrocellulose or nylon membrane by a process called Southern blotting (5).

The DNA in the gel is first denatured (separated into single strands) by immersion in alkali, then neutralized in a buffered solution. The gel is then placed on a filter-paper wick which is fed by a reservoir of salt solution. The membrane on to which the DNA is to be transferred is placed on top of the gel under a stack of tissue or filter paper and several heavy weights. The DNA is drawn up on to the membrane as the salt solution is absorbed up through the gel and into the stack of tissue paper (*Figure 2.5*). The DNA is then fixed to the membrane by baking in an oven at 80°C.

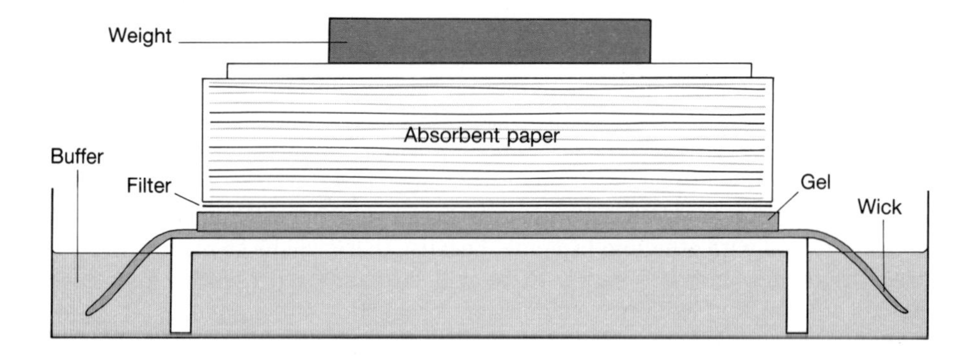

Figure 2.5. Southern blotting to transfer DNA from a gel to a membrane (filter).

4. Hybridization with labelled DNA

Single-stranded DNA bound to a membrane can hybridize with DNA fragments in solution. Therefore, a sequence of interest can be isolated, labelled, and used to visualize similar sequences bound to the filter.

There are a number of methods by which the 'probe' DNA can be labelled. Probes vary in length from very short 'oligonucleotide' sequences (usually less than 50 bp long) up to several kilobases. Oligonucleotides are generally labelled using the enzyme T4 polynucleotide kinase. In an ATP-dependent process, this enzyme catalyses the phosphorylation of the 5′-hydroxyl group of deoxyribose. The terminal (gamma) phosphate group of the ATP is donated to the terminus of the DNA. If the gamma phosphate group of ATP has been made radioactive with [^{32}P]-phosphorus, then the 5′ end of the oligonucleotide sequence will become isotopically labelled (*Figure 2.6*).

Longer sequences of double-stranded DNA are usually labelled by either 'nick translation' or 'random priming'. The first step in making a probe by nick translation is to treat the DNA with a very small quantity of the endonuclease DNAse I, which produces single, random nicks in the strands of DNA. Then a sufficient quantity of each nucleotide is added, including at least one which is labelled with ^{32}P in the alpha phosphate group, together with the enzyme DNA polymerase I. This enzyme has both polymerase and 5′-to-3′ exonuclease activity, so, beginning at the nicks, it digests away the existing unlabelled strand,

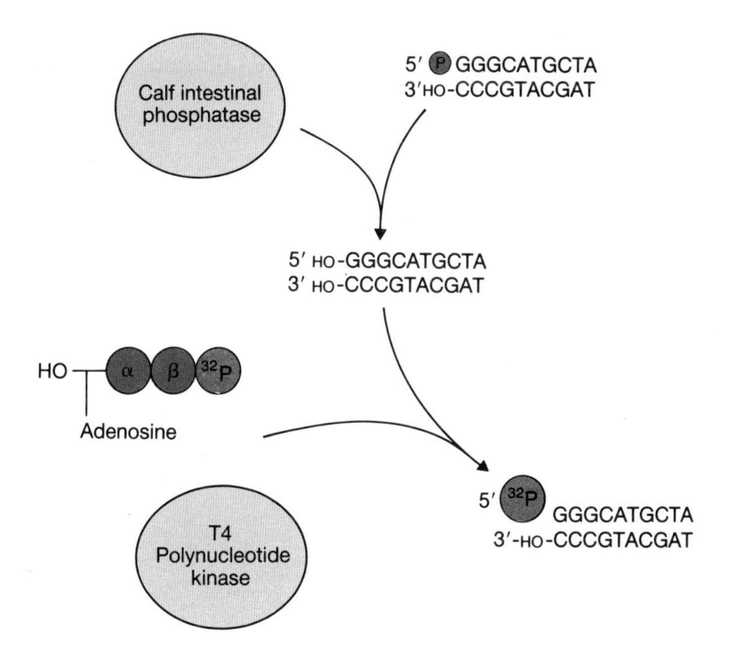

Figure 2.6. End-labelling of DNA using T4 polynucleotide kinase.

and simultaneously synthesizes a new strand incorporating the radioactive nucleotide (*Figure 2.7*).

The process of random priming uses a version of DNA polymerase I which has had the site for the 5′-to-3′ exonuclease activity cleaved off. This enzyme is known as the Klenow fragment. The first step during random priming is to denature the probe DNA, which is accomplished either by boiling or adding alkali. Then a mixture of random oligonucleotides is added and annealed to the template DNA. The Klenow fragment will then incorporate nucleotides (including the radioactively labelled nucleotide) to make double-stranded copy from the short, double-stranded templates where the random hexamers have annealed (*Figure 2.8*). A hexamer will anneal in a random sequence of DNA approximately once every 4000 bp (assuming a 50:50 ratio of GCs and ATs).

Prior to adding a labelled double-stranded DNA probe to the hybridization solution, it must be denatured (usually by boiling). During the course of the hybridization reaction with membrane-bound DNA, the probe DNA will be re-annealing in solution. This competing reaction can be eliminated by producing single-stranded probes. This is usually accomplished by subcloning the probe sequence into the replicative form (double-stranded) phase of the single-stranded

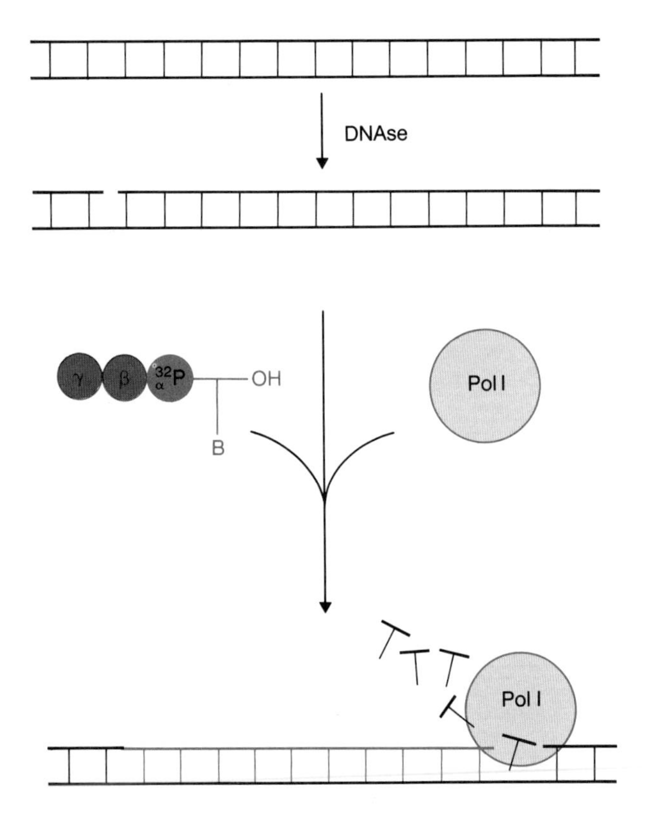

Figure 2.7. Labelling of DNA by nick translation.

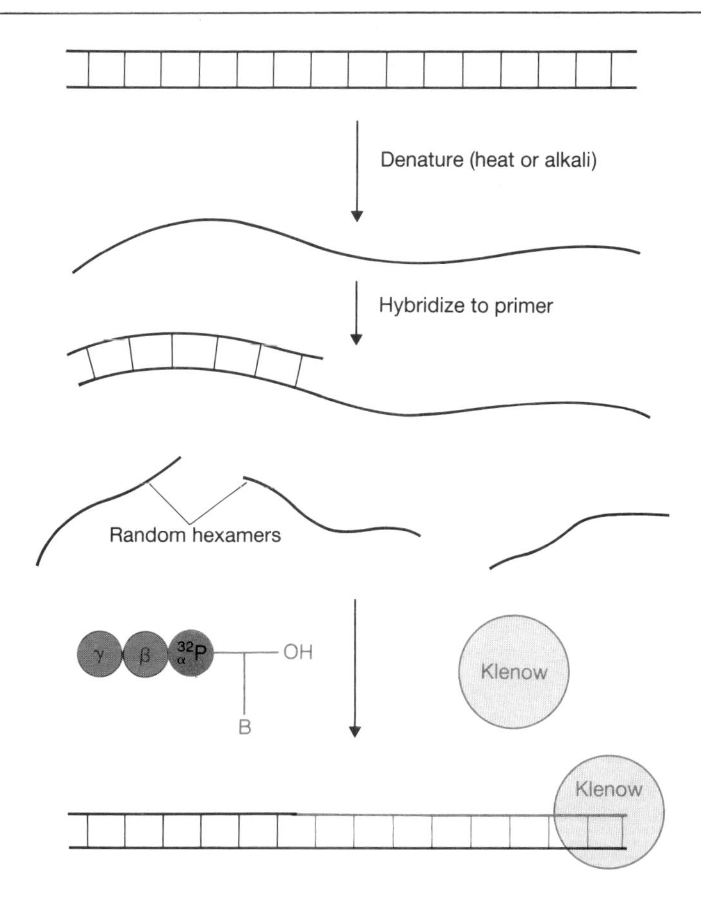

Figure 2.8. Labelling of DNA by random priming.

phage, M13. The Klenow fragment is then used to produce a labelled copy using the 'universal' primer site in M13. If the production of double-stranded copy is timed so that the enzyme only reads across the required sequence, then after denaturing the new radioactive strand from the M13 molecule, the labelled single-stranded probe can be isolated in an agarose gel (which will separate the fragments according to size).

Hybridization is carried out in a salt solution. The dynamics of the reaction are dependent on salt concentration and on temperature. High temperature and low salt concentration produce very stringent conditions, so that only sequences with perfect homology will hybridize. Unwanted background absorption of DNA to the membrane is minimized by adding a protein to block non-specific binding with blank areas of the membrane. Degraded non-homologous DNA is also often added as a competitor to minimize partial hybridization of the probe to non-target DNA sequences. After hybridization, the membrane is washed in a salt solution to remove excess and partially bound probe. The filter is then wrapped in cellophane while still damp, and exposed for hours or days to X-ray film. The sequences of interest will show up on the X-ray film as black bands.

5. Isolation and restriction analysis of mtDNA

The DNA in mitochondria is a closed circular molecule of about 16 kb in animal species, and generally larger in plants. The simplest way to extract mtDNA is first to isolate whole mitochondria. This is accomplished by gently grinding cellular material in a glass homogenizer so that the cell walls are broken down, but not the organelles. The homogenized preparation is then centrifuged slowly (1000 *g* for 10 minutes) so that cellular material and nuclei are pelleted at the bottom of the tube. The supernatant is then pipetted off and spun again, but much faster (10 000 *g* for 10 minutes). This time the mitochondria are pelleted (*Figure 2.9*). These are lysed with a detergent solution and the mtDNA extracted with phenol (see 6).

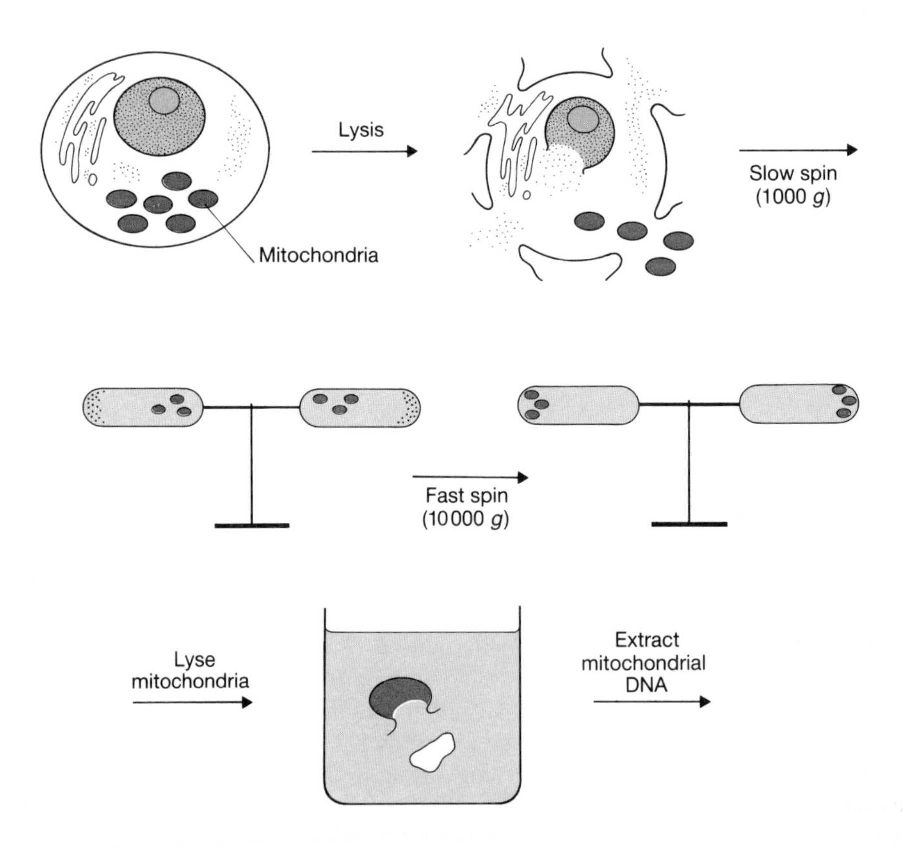

Figure 2.9. Isolation of mitochondrial DNA by differential centrifugation. Further purification by sucrose or caesium chloride gradient centrifugation is often required.

At this stage there is usually contaminating nuclear DNA in the preparation. The closed circular (mtDNA) molecules can then be separated by spinning the DNA at very high speed (approximately 200 000 *g* or more) in an extremely dense salt solution (caesium chloride). When the salt concentration is correctly adjusted, RNA will pellet at the bottom of the tube, and nuclear and mtDNA will form separate bands towards the centre of the tube.

The purified mtDNA can then be cleaved with restriction enzymes and separated on an agarose or polyacrylamide gel (see Section 3). Polyacrylamide is a better medium for discriminating small DNA fragments. The fragments can be visualized with either ethidium bromide or silver staining; the latter is far more sensitive. A high level of sensitivity can also be achieved by radioactively end-labelling the fragments with kinase. The procedure is the same as that described for labelling oligonucleotides (see Section 4).

Since most mitochondrial genomes are circular, a restriction enzyme that cuts the mtDNA *n* times will produce *n* fragments. If it is assumed that all nucleotides are randomly distributed in the genome, then the expected frequency of a given restriction site can be estimated by

$$a = (\tfrac{1}{2}g)^{r_1}[\tfrac{1}{2}(1-g)]^{r_2}$$

where r_1 is the number of guanines (G) and cytosines (C), and r_2 the number of adenines (A) and thymines (T) in the restriction site and g is the percentage G+C content of the genome (7). For example, if g has a value of 0.5 and the size of the mitochondrial genome is 16 000 bp, then the expected frequency of restriction sites for the enzyme *Eco*RI (which cuts at GAATTC) will be 0.03 per cent (producing 3.9 restriction fragments). The enzyme *Alu*I, which cuts at the four-base sequence AGCT, should produce 62.5 fragments. The actual number of fragments will, of course, vary, but it is expected to be of this order of magnitude.

6. Polymerase chain reaction

The polymerase chain reaction (PCR) allows the enzymatic amplification of microgram quantities of specific DNA sequences. This technique has greatly facilitated the analysis of sequence variation, and has enabled a new level of phylogenetic investigation. Knowledge of the DNA sequence for a particular region from a few taxa can permit the amplification of the same sequence from quite distantly related species by this method. It can also be a very fast way to screen for variation between a large number of individuals in a study population in combination with sequencing or RFLP analysis.

The principle behind PCR is very simple. Short oligonucleotide 'primers' are designed so that they will anneal on either side of a 'target' sequence. This sequence can be any size from about 100 bp to 10 kb. The target sequence is denatured, the primers annealed, and a copy is produced from each strand using a polymerase. Then the cycle of denaturing, annealing, and extension is repeated

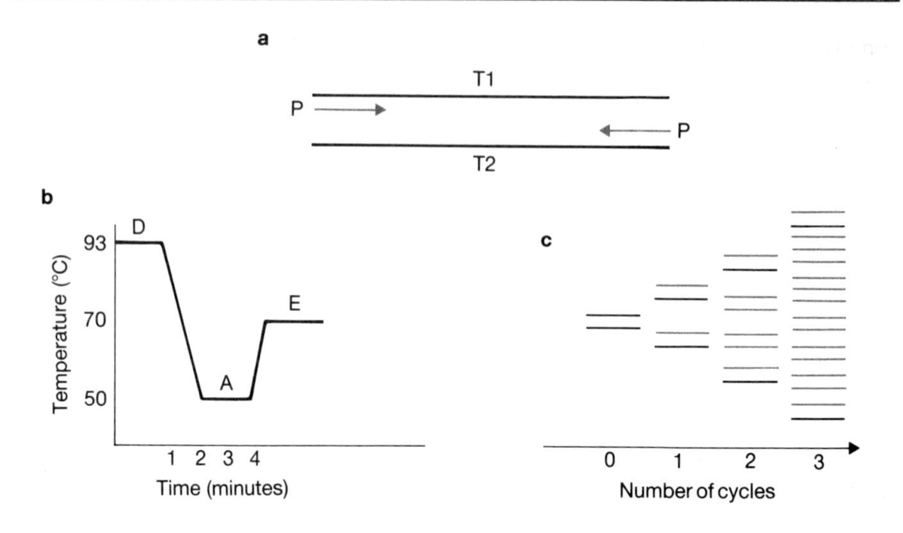

Figure 2.10. The polymerase chain reaction (PCR). **a** Primers anneal to denatured template DNA. **b** The reaction is cycled between denaturing, annealing and extension temperatures. **c** Repeated cycling results in an exponential amplification of the target sequence.

so that copies are now made from both original and copy templates. The result after numerous repeated cycles is an exponential amplification of the target sequence (*Figure 2.10*).

The first experiments with PCR used the Klenow fragment to polymerize during the extension phase (8). However, Klenow activity is denatured by the high temperatures necessary for denaturing the template DNA. Therefore, more enzyme must be added at the start of each cycle. This was both labour-intensive and expensive. The technique was revolutionized by the discovery of a heat-stable polymerase (isolated from thermophilic bacteria which live in thermal pools) called '*Taq* polymerase' (9). This enzyme can be added just once at the beginning of the experiment. The development of programmable thermocycling machines and the use of *Taq* polymerase has made PCR a routine procedure in many molecular biology laboratories.

7. DNA sequencing

DNA sequencing provides the maximum resolution for computing genetic distance (see Chapter 3). Further, a comparison of the sequence under study from a diverse range of taxa will help determine the dominant mode of change affecting the evolution of that sequence. Patterns and repetitive motifs can be indicative of specific DNA turnover mechanisms (see Chapter 1). For example, significant levels of simplicity (repetitious elements) often indicate the process of DNA slippage (Chapter 1, Section 5.3.3). Clearly, it is important to know something about the mode and rate of change before attempting time-dependent interpretations, such as genetic distance.

There are two general methods for determining the sequence of nucleic acids in a segment of DNA: the 'chemical cleavage' procedure (usually referred to as the Maxam and Gilbert method; see 10) and the 'chain termination' procedure (11). In the Maxam and Gilbert method the DNA to be analysed is cut with a restriction enzyme, and the fragments are isotopically end-labelled (see Section 3). The fragments are gel isolated, and a fragment to be sequenced is cleaved with another restriction enzyme producing two fragments, each isotopically labelled at one end. These fragments are isolated on a gel, and treated with chemicals that cause partial random modification of the DNA in a base-specific manner. In a subsequent chemical reaction the DNA will be cleaved where it has been modified. For example, one in every few hundred guanine residues is methylated by dimethyl sulphate. Piperidine can then be used to cleave the DNA at the methylated sites and generate a pool of molecules cleaved at different G residues. When this is run on a vertical polyacrylamide gel, a ladder of bands shows the size of each fragment. The resolution on this type of gel is good enough

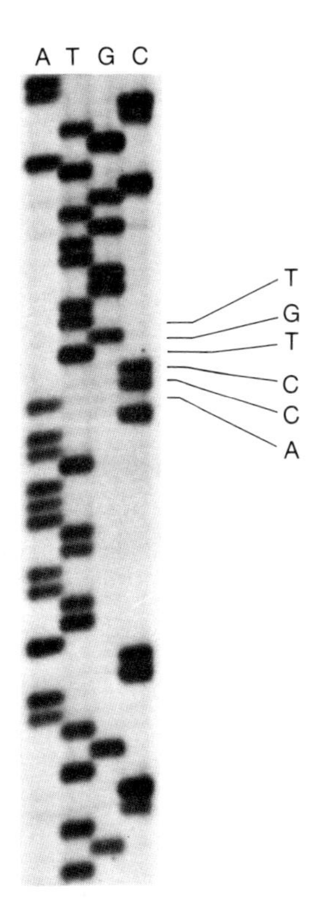

Figure 2.11. DNA sequencing. Reactions from the chain termination procedure run on a 6 per cent polyacrylamide gel. A portion of the sequence indicated by the four reactions is given at the right of the gel.

to distinguish even single base-pair differences. This type of combination of chemical reactions is done for all four bases, and the reactions run side by side.

The chain termination method is based on the premature termination of synthetic strands, and in this way is effectively the reverse of the Maxam and Gilbert procedure. The template must be single-stranded (either cloned into M13 or denatured prior to labelling). A suitable oligonucleotide is used to prime the polymerase reaction, and the preparation is divided four ways. Each reaction has all deoxynucleotides (A,T,G, and C), including one that has been isotopically labelled, and a smaller quantity of one dideoxynucleotide (either A,T,G, or C), which lacks the 3′-OH group necessary for DNA chain elongation. Each reaction produces a pool of labelled fragments terminated at random at one of the four bases. As for the Maxam and Gilbert method, the four reactions are run side by side on a vertical polyacrylamide gel (*Figure 2.11*). The polymerase used can be the Klenow fragment, but is most often a version called 'Sequenase' which is less susceptible to false terminations than the Klenow enzyme. Radioactive sulphur (^{35}S) is usually used to label one of the nucleotides instead of ^{32}P, because ^{35}S is much less energetic and so provides substantially better resolution of individual bands. The chain termination method represents a considerable improvement over the Maxam and Gilbert method, saving effort and improving quality and range, and is now the most commonly used sequencing protocol (although Maxam and Gilbert is still useful for some applications). By the chain termination method, up to 300–400 bp can be read from one reaction sequencing from double-stranded DNA, and over 500 bp sequencing from single-stranded DNA.

8. Further reading

Protein analysis

Harris,H. and Hopkinson,D.A. (1978) *Handbook of Enzyme Electrophoresis in Human Genetics*. American Elsevier, New York.

Hames,B.D. and Rickwood,D. (1990) *Gel Electrophoresis of Proteins—a Practical Approach*, 2nd edn. IRL Press at Oxford University Press, Oxford.

Molecular techniques

Perbal,B. (1988) *A Practical Guide to Molecular Cloning*. Wiley, New York.

PCR

Erlich,H.A. (ed.) (1989) *PCR Technology*. Stockton Press, New York.

DNA sequencing

Howe,C.J. and Ward,E.S. (ed.) (1989) *Nucleic Acids Sequencing—a Practical Approach*. IRL Press, Oxford.

9. References

1. Hames,B.D. and Rickwood,D. (1981) *Gel Electrophoresis of Proteins—a Practical Approach*, 2nd edn. IRL Press at Oxford University Press, Oxford.
2. Brewer,J.G. (1970) *An Introduction to Isozyme Techniques*. Academic Press, New York.
3. Selander,R.K., Smith,M.H., Yang,S.V., Johnson,W.E., and Gentry,J.B. (1971) *Stud. Genet. Univ. Texas Publ.*, **6**, 49.
4. Harris,H. and Hopkinson,D.A. (1978) *Handbook of Enzyme Electrophoresis in Human Genetics*. American Elsevier, New York.
5. Southern,E. (1975) *J. Mol. Biol.*, **98**, 503.
6. Lansman,R.A., Shade,R.O., Shapira,J.F., and Avise,J.C. (1981) *J. Mol. Evol.*, **17**, 214.
7. Nei,M. and Li,W.-H. (1979) *Proc. Nat. Acad. Sci. USA*, **76**, 5,269.
8. Mullis,K.B. and Faloona,F. (1987) *Meth. Enzymol.*, **155**, 335.
9. Saiki,R.K., Gefland,D.H., Stoffel,S., Scharf,S., Higuchi,R.H., Horn,G.T., Mullis,K.B., and Erlich,H.A. (1988) *Science*, **239**, 487.
10. Maxam,A.M. and Gilbert,W. (1977) *Proc. Nat. Acad. Sci. USA*, **74**, 560.
11. Sanger,F., Nicklen,S., and Coulson,A.R. (1977) *Proc. Nat. Acad. Sci. USA*, **74**, 5,463.

3

Statistical interpretation of variation and genetic distance

1. Genetic diversity

The analysis of variation at the molecular level began at the turn of the century with studies on blood types in humans (1). However, the more extensive characterization of variation in other species did not begin until 1966 with the application of gel electrophoresis (2,3). In the first 10 years after the technique was introduced, genetic variation at loci coding for proteins was described for nearly 250 species (where 14 loci or more were investigated; see 4). It became apparent that there is extensive genetic variation in natural populations. To date well over 1000 species have been investigated.

Isozyme variation in the species or population under study is usually described in terms of the proportion of polymorphic loci per population (*P*) and heterozygosity (*H* denotes heterozygosity per locus per individual). At a gene locus with two alleles (variants, *A* and *a*; possible genotypes, *AA*, *Aa*, and *aa*), assuming random mating, the allele frequency is defined below. If the total number of individuals in the population is *N* let the number possessing each genotype be N_{AA}, N_{Aa}, and N_{aa}. The frequency *p* of the *A* allele in the population will be

$$p = 2N_{AA} + N_{Aa}/2N$$

and the frequency *q* of the *a* allele will be

$$q = 2N_{aa} + N_{Aa}/2N.$$

The heterozygosity (*H*) is the proportion of *Aa* genotypes, which according to the Hardy–Weinberg rule is $2pq$ at equilibrium ($p^2 + 2pq + q^2 = 1$). This rule states that gene and genotype frequencies will remain constant between generations in an ideal population (an infinitely large random mating population, obeying the Mendelian rules of segregation where there is no selection, migration, or mutation), that is, that meiosis and mating behaviour do not alter gene

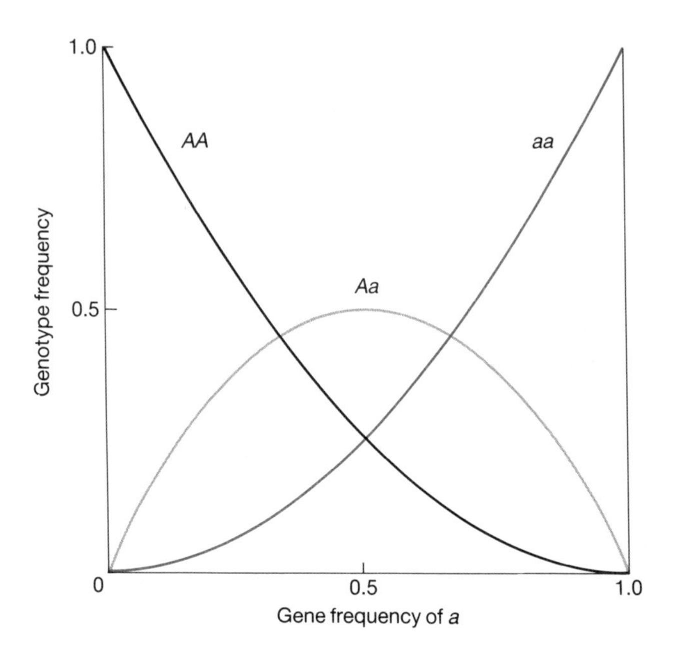

Figure 3.1. Relationship between genotype frequency and gene frequency for two alleles in a population that is in Hardy–Weinberg equilibrium (after 30).

frequencies (*Figure 3.1*). If in a population of 1000 there were 250 *AA*, 500 *Aa*, and 250 *aa* individuals, then both p and q would be 0.5, and the population could be said to be in Hardy–Weinberg equilibrium.

In multiple-allele systems heterozygosity can be defined as

$$1 - \sum x_i^2$$

where x_i is the frequency of the ith allele at a given locus. H for a species is simply the average heterozygosity for all loci investigated. In non-random mating populations the above quantity is not related to the frequency of heterozygotes, but is nevertheless a good measure of genetic diversity. Note that this same formula can be used to describe the heterozygosity of restriction sites in DNA (RFLP data). When DNA sequence data are compared, the 'nucleotide diversity' (heterozygosity at the nucleotide level) is defined by Nei (5) as

$$\pi = \sum x_i x_j p$$

where x_i is the population frequency of the ith DNA sequence and p is the proportion of different nucleotides between the ith and jth sequences. This can be estimated by

$$\pi = \sum p/n_c,$$

where n_c is the total number of comparisons [if n is the number of sequences in the sample, then $n_c = \frac{1}{2} n (n-1)$]. For example, if 10 sequences (individuals) each 100 bp long are compared and only one is different, at one base-pair, then the nucleotide diversity is 0.003. The nucleotide diversity values for vertebrate mtDNAs range from 0.004 (humans) to 0.013 (chimpanzees) (5).

Nevo and co-workers (6) computed average values for protein polymorphism (P) and heterozygosity (H) based on 968 plant and animal species. They found values of 0.284 ± SD 0.197 and 0.073 ± SD 0.076 respectively. This means that on average 28.4% per cent of the loci of each species are polymorphic, and the average heterozygosity (including monomorphic loci in the calculation) is 7.3 per cent. The average for 551 vertebrate species was lower (less variation): $P = 0.226 \pm$ SD 0.146 and $H = 0.054 \pm$ SD 0.059. Mammals were at the low end of that group: $P = 0.191 \pm$ SD 0.137 and $H = 0.041 \pm$ SD 0.035.

Caution is necessary in the interpretation of average P values for protein data, however; sample sizes vary greatly between studies, as do criteria for polymorphism. A locus is most commonly defined as polymorphic if the most common allele frequency is 0.99 or less, but other criteria are used and not always stated in the published report. Further, when a small number of loci are investigated (e.g. 24 or less; see 7) the estimate of average heterozygosity is subject to a large standard error. In the review published by Nevo (4), 24 loci or less were investigated in 74 per cent of the studies. Nei (8) recommends that estimates of average heterozygosity be conducted on as many loci as possible, ideally a random sample of the genome. The number of individuals on the other hand, can be as low as 20 (9).

Selection will affect the expression of new variation through the differential survival of favourable and deleterious phenotypes. Selection can also maintain variation if the heterozygous condition is favoured at a given locus or through various mechanisms of 'balanced' selection. However, this type of selection can only act directly on DNA sequences that are expressed phenotypically.

2. Interpopulation diversity

Nei (8) describes the analysis of gene diversity within subdivided populations. These measures are related to genotype frequencies only in random mating populations. Subdivision itself can affect genotype frequencies. If a species is divided into subpopulations where there is random mating, and if gene frequencies differ from subpopulation to subpopulation, then for the species as a whole, homozygote genotypes will increase at the expense of heterozygotes (10). This is known as the Wahlund effect, and it has the same effect on overall heterozygosity as inbreeding.

Polymorphisms at a given set of loci can provide only a statistical distinction between populations. That is, although the means for the two populations may differ, there is generally considerable overlap between the two distributions of multilocus genotypes. Given some *a priori* criteria for separating populations and sufficient variation, it is possible to use genetic diversity measures to distinguish between breeding stocks. However, when sorting stocks from a mixed assemblage, it is not possible to use allele frequencies to classify individuals chosen at random, although a maximum likelihood method can be used to estimate the composition of the mixture. One solution is to look for a unique allele, or genetic 'marker', that is indicative of a given population.

Nei's measure of gene differentiation between populations, G_{ST} (see Appendix *3.1*), can be regarded as an extension of Wright's correlation between two gametes drawn at random from each subpopulation, the *F*-statistic: F_{ST}. This is an estimation of the fixation index from a group of sample populations and is given by the actual gene frequency variance divided by the limiting variance for each allele. However, Wright's (11) application of the *F*-statistic was devised in terms of neutral genes. Further, there is an assumption that the number of subpopulations is infinitely large. The principle distinction between Wright's and Nei's formulations is that Wright defines the *F*-statistic as a correlation between uniting gametes, while Nei compares observed and expected heterozygosities. Nei (12) redefines the application of *F*-statistics to this problem as a function of heterozygosities. So defined, it is independent of the number of subpopulations or alleles involved and can be applied whether or not there is selection. G_{ST} and F_{ST} are not related to the frequency of heterozygotes except in populations that are consistent with the Hardy–Weinberg rule.

G_{ST} and the degree of gene differentiation, D_m (see Appendix *3.1*) have been applied to numerous studies on protein polymorphisms in regional populations and races. For example, Nei and Roychoudhury (13) looked at 62 protein loci in the three principle human races (caucasoid, negroid, and mongoloid). The minimum net codon differences between the three races were estimated to be 0.0195 per locus (D_m). G_{ST} was estimated to be 0.088, which means that 8.8 per cent of the total gene diversity can be attributed to genetic differences between the races. This indicates that variation within race accounts for most of the genetic diversity.

3. Genetic distance

Nei (14) defines genetic distance as a measure of gene diversity between populations expressed as a function of genotype frequency. It is a statistical measure providing a standardized scale for the quantification of genetic differences. A variety of measures have been devised, but the one most commonly used for comparing protein polymorphism is Nei's (14) standard genetic distance. A simple measure of genetic similarity (I) is multiplied by the natural logarithm (ln) to give a parameter that is 0.0 for genotypes that are completely dissimilar (*Figure 3.2*). If individual codon changes are independent, the mean number of net codon differences, the standard genetic distance, is given by

$$D = \ln(I),$$

I being given by

$$I = J_{XY}/J_XJ_Y$$

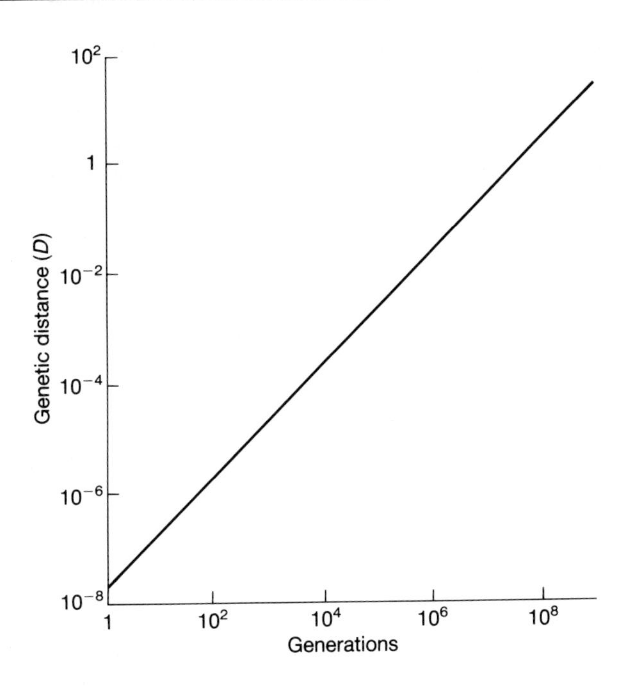

Figure 3.2. Relationship between Nei's genetic distance measure (D) and divergence time in generations (after 5).

where J_{XY} is the mean probability (taken over all loci) that two genes chosen at random, one from each of two populations X and Y, are identical. J_X and J_Y represent the mean probabilities that two genes chosen at random from a single population (X or Y respectively) are identical. For example, if J_{XY} is 0.95 and J_X and J_Y each have the value of 0.98, then I is 0.96 and D is 0.011. This is within the range expected for differences between local animal populations within a species: 0.001 (lizards) to 0.049 (*Drosophila*) (5).

If the rates of codon changes vary from locus to locus (as is the case; see Chapter 1), D will be an underestimate. In this case genetic distance can be estimated with the same formulation as for D, except that J_{XY}, J_X and J_Y are computed as the geometric means. This is designated the maximum genetic distance, D' (see examples in 8). Sampling errors of gene frequencies can greatly inflate this measure, however, and if there is even one locus where there is no common allele between two populations, D' will be infinitely large.

For comparing restriction site polymorphisms, a simple measure of similarity is the proportion of fragments shared between the restriction enzyme digestion profiles. This is given by:

$$F = 2n_{XY}/(n_X + n_Y)$$

where n_X and n_Y are the number of fragments for populations (or individuals) X and Y, and n_{XY} is the number of shared fragments (15,16). Upholt (15) has related this quantity to an estimate of the number of base substitutions per nucleotide separating two populations (p):

$$p = 1 - \{ [(F^2 + 8F)^{1/2} - F]/2 \}^{1/n}$$

where n is the number of base pairs recognized by the restriction enzyme. Nei and Li (16) derive the following estimate for this measure:

$$d = -(\ln F)/n.$$

This is based on the assumption that F can be used as an estimate of the proportion of ancestral restriction sites that have remained unchanged in both populations. These two formulations produce very similar results. This measure is most accurate when F is large, as the variance of d becomes very large when F is small (5). Enzymes that recognize four base pairs would be expected to reveal greater variation than enzymes that recognize six base pairs. For example, if n is 6 and F is 0.90, then d is 0.018. However, if n is 4 then F would need to be 0.93 for the same distance value.

When possible, it is preferable to compare actual shared restriction sites. This can be accomplished by mapping the sites in the region under study through a series of double digests. For example, if two enyzmes (A and B) each cut the DNA once, then digesting the DNA with both will show which of the two fragments left by enzyme A is cut by enzyme B.

Both formulations are based on a number of assumptions:

1. *Nucleotides are randomly distributed in the genome.* Although this is not the case, Nei and Li (16) suggest that small deviations from randomness will not significantly alter the results. However, regions of DNA that have been affected by DNA turnover mechanisms that generate repeated motifs (see Chapter 1) could be expected to show a substantial deviation from the expected number of restriction sites.

2. *Fragment variation arises solely by base substitution.* This will depend on the region of DNA under study. For example, the mitochondrial genome is an extremely useful genetic component for comparisons of population level variation using RFLP data (see Chapter 4, Section 3), and is thought to change primarily through the accumulation of base substitutions.

3. *Nucleotide substitution rates are the same for all nucleotides.* This is seldom the case. As discussed in Chapter 1, nucleotide substitution rates vary by several orders of magnitude. However, Nei and Chakraborty (17) point out that when the number of nucleotide differences per nucleotide site is small (as for intraspecific studies), this assumption does not produce any serious errors. If the distance measure is large (say more than 0.3), then it will be an underestimation (16).

4. *All restriction fragments can be detected, and fragments of similar length are not scored as identical.* Enzymes which digest the genome into relatively few fragments (enzymes where n is 5 or 6) can be used to avoid this problem. However, Nei

and Tajima (18) point out that the accuracy of the distance measure is enhanced by using enzymes that produce large numbers of fragments, and suggest that the results will not be greatly affected if a few small fragments are not detected. Double digests, as described above, will help sort out co-migrating fragments.

Many of the difficulties associated with estimating genetic distance based on enzyme or RFLP data can be surmounted by the direct comparison of DNA sequences. In this way local differences in the rate of change can be quantified, and biases in the composition of sequences (G + C content) can be assessed. The genetic distance between homologous sequences is usually either represented simply as the percentage difference, or corrected for multiple substitutions at a given site as formulated by Kimura and Ohta (19):

$$d = \frac{3}{4} \ln(1 - \frac{4}{3}k)$$

where k is the percentage difference in base composition. When k is small the correction is very slight. For example, if k is 10 per cent d equals 0.1002 and when k is 25 per cent d equals 0.281.

When comparing two populations, the net difference between them can be estimated by subtracting the average level of variation within populations from the level of variation between them:

$$d = d_{XY} - \frac{1}{2}(d_X + d_Y),$$

where d_{XY} is the average from pairwise comparisons between populations, and d_X and d_Y are the averages from pairwise comparisons within populations X and Y, respectively. For example, if the pairwise distance between two populations was 10 per cent, but there was 2 per cent variation within one population, and 4 per cent within the other, then the corrected interpopulation distance measure would be 7 per cent (by the formula given above).

Nei (8) has applied his measure of standard genetic distance ($D = -\ln I$) to a number of studies on populations within species, subspecies, and higher taxonomic divisions. Genetic distance between races (or populations) was always less than a few per cent. Seven studies comparing subspecies showed a range of D values from 0.004 to 0.351. At all levels there is considerable variation in the estimates of D. However, on average, higher taxonomic divisions had higher values.

For comparisons between populations that are in equilibrium with respect to random genetic drift, mutation and selection, there is a simple relationship between genetic distance and divergence time, t. For nuclear genomes t is equal to $D/2a$, and for mitochondrial genomes t is equal to D/a, where t is the divergence time in millions of years, a is the rate of substitution per year per unit of study (locus, mitochondrial genome, or haplotype), and D is genetic distance expressed as a percentage. This assumes that substitutions accumulate at a constant rate. This may not be true for a given sequence because of the influence of molecular turnover (see Chapter 1). Further, demographic factors (such as extreme fluctuations in population size or a population bottleneck) may have reduced variation at some point in the past.

4. Variance of heterozygosity and genetic distance

When planning a study of heterogeneity and genetic distance between sample populations, the number of loci and individuals that should be investigated can be approximated by minimizing the sampling variance (7):

$$V(H) = [V_g(h) + V_s(h)]/r.$$

If the total number of genes to be studied is held constant (r_n), then clearly the variance at locus H, $V(H)$ (Appendix *3.2*) can be minimized by maximizing r (the number of loci studied). Nei and Roychoudhury (7) illustrate this point by analysing data on three cave and nine surface populations of a characid fish species. Sample variances were computed for comparisons of populations of different sizes and for different numbers of loci. For the purpose of estimating average heterozygosity or genetic distance, they suggest that as few as 20 individuals per locus should be sufficient as long as a large number of loci (say 30 to 70) are investigated. However, if the number of individuals is too small, the bias of the heterozygosity estimate becomes large. The general conclusions were that few individuals need be studied when genetic distance is fairly high, and when heterozygosity levels are low (providing numerous loci were analysed). Conversely, and when the number of loci that can be investigated is limiting, then a larger number of individuals will improve the precision of the results.

5. Effect of migration on population diversity

Surprisingly little mixing is necessary to overcome the effects of genetic drift and to maintain genetic homogeneity between populations. Crow and Kimura (20) describe the conditions for the establishment of an equilibrium between migration and random drift. In a random mating population the probability that two gametes will have identical genes is $1/2N$. The chance that two gametes have different parental genes is $1-(1/2N)$ ($2N$ is the number of genes at a given locus in a population of N diploid parents). The inbreeding coefficient is given by

$$f_t = \frac{1}{2N} + \left(1-\frac{1}{2N}\right)f_{t-1}$$

where f_{t-1} is the inbreeding coefficient for an average individual in the previous generation. If we consider a group of subpopulations with a migration rate between them of M, the probability that neither of the two uniting genes will be displaced by a migrant gene is $(1-M)^2$. The increase in autozygosity of a subpopulation is given by

$$f_t = \left[\frac{1}{2N_e} + \left(1-\frac{1}{2N_e}\right)f_{t-1}\right](1-M)^2$$

where N_e is the effective (reproducing) number in the subpopulation. At equilibrium f_t is equal to f_{t-1} which is equal to f, and when M is small M^2 can be neglected, so that

$$f = \frac{1}{4N_eM} + 1$$

This means that if M is very much less than $1/4N_e$, then f will be large and the populations will tend to diverge. However, if M is larger than $1/4N_e$, then the subpopulations are effectively a single panmictic unit. In practical terms, if one or more reproductively active individual migrates between subpopulations per generation, then there will be little local differentiation. The effect will be less pronounced in species that tend to disperse over a short range, because neighbouring subpopulations will tend to be genetically similar. Further differentiation could result from selection pressure (see below) or a variety of DNA turnover mechanisms (see Chapter 1). In this case a higher level of exchange would be necessary to eliminate genetic differentiation between populations.

Considering only the effects of genetic drift, Nei and Feldman (21) develop a formulation for the effect of migration on the normalized identity of genes between two populations. If $M = m_1 + m_2$, where m_1 and m_2 stand for the migration rates from populations 1 and 2 respectively, then

$$I = M/(M + 2u)$$

where u is the mutation rate per locus per generation. This quantity $2u$ is very small, so I will be very nearly unity unless M is very small. This means that genetic distance between populations cannot be large unless migration rates are very low.

Given a computed measure of genetic distance between two populations, it is possible to estimate the maximum possible migration rate that could have occurred (8). Assuming that the genetic distance between the populations has reached a steady-state value, then I equals $\exp(-D)$ which equals $m/(m + u)$, where m is the maximum migration rate and u is the mutation rate per locus per generation. Therefore,

$$m = u \exp(-D)/1 - \exp(-D).$$

Comparing human races, assuming a mutation rate of 2×10^{-6} per gene per generation, Nei (8) derives a maximum migration rate of 1×10^{-4} per generation between caucasians and negroids, and 2×10^{-4} between caucasians and mongoloids. This suggests that there are less than 1 in 10 000 mixed racial couples in human populations. However, it should be noted that this measure gives a long-term average estimate, and assumes an equilibrium population structure, constant migration rates, and equal numbers of individuals in population subdivisions. It is most useful as a relative, rather than as an absolute measure.

Wright (11) has discussed the interaction between migration and selection for his island model. Subpopulations are thought of as being isolated, but with limited exchange of individuals. By this model the populations will differentiate only when the selective force is much larger than the influence due to migration. When selective and migration forces are equal, a gene that is favoured by local selection will have a frequency equal to the square root of the average allele frequency for the entire population.

6. Effective population size

The effective population size (N_e) is the size of an ideal population that would show the same genetic characteristics as the real population (see Chapter 1, Section 2 for an alternative definition). There are three general ways that genetics can be applied to the estimation of effective population size. First, a long-term estimate of N_e can be derived based on the theory that, for a population under selective neutrality, heterozygosity at equilibrium is a function of N_e and the neutral mutation rate. The general relationship is as follows:

$$N_e \approx \frac{H}{4\mu(1-H)}$$

where μ is the neutral mutation rate. When recombination is absent or very rare, as is the case with vertebrate mtDNA (see Chapter 4, Section 3), then this can be estimated as follows (after 22):

$$N_e = 10^8\pi/sg$$

where π is the nucleotide diversity, s is the per cent substitution rate per million years, and g is the generation time in years. For example, π for human mtDNA is about 0.004 (22), generation time is about 20 years, and the mtDNA substitution rate is about 2 per cent per million years (22). Therefore an estimated N_e for humans would be 10 000 (see Chapter 4, Section 3.4 for further discussion). This method is most suitable for giving a general, long-term estimate of N_e for relative comparisons. It is important that the substitution rate for the particular region of DNA under analysis is known, as rates vary in different regions of both the nuclear and mitochondrial genomes.

A second method uses short-term changes in genetic parameters to estimate current N_e. Allele frequencies and the degree to which loci are linked (close together on a chromosome) are both affected by N_e. Therefore, by measuring the temporal fluctuations in allele frequencies between generations (23), or the levels of linkage disequilibrium at multiple loci (24), current N_e can be indirectly estimated. These methods depend on the assumptions of selective neutrality, random mating, and no migration between genetically differentiated populations. To reduce variance to acceptable levels it is necessary to compare a large number of generations (for the temporal allele method), or a large number of loci (for the linkage disequilibrium method). A detailed description of the mathematical formulations and precise requirements for these methods is beyond the scope of this book (but see 23 and 24).

A third application for genetic techniques would be as a means of determining accurate information to be used in a demographic model. For example, N_e depends on the number of effective breeders in a population. If the number of active breeders in one sex is limiting, then the effective size of the population will be limited by that sex:

$$N_e = 4(N_m N_f)/(N_m + N_f)$$

where N_m and N_f are the effective number of males and females respectively. Determining N_e this way requires knowledge of variance in reproductive success for each sex, but this can be very hard to determine for male animals. A way around this is to establish genealogies by paternity testing. Genetic methods for paternity testing are described in the next section.

7. Kinship assessment

Tests for paternity using enzyme data of nuclear-encoded genes are possible by exclusion analysis. If a female and her offspring can be identified and their allelic patterns determined, then potential fathers not possessing the necessary alleles can be eliminated as possible fathers. Various statistical methods have been employed to calculate the probability of 'non-paternity' and the 'likelihood of paternity' from allozyme data. It is also possible to identify a skew in male reproductive success by examining whether the distribution of paternal allele frequencies in offspring differs from allele frequencies in the adult male population (25).

The use of allozyme variation for paternity testing has been largely replaced by the analysis of 'minisatellite' loci by the procedure known as DNA 'fingerprinting' (26). Exclusion by comparing allozyme genotypes is useful as an initial screen when the sample size is large. The discovery and nature of minisatellites is described in more detail in Section 6 of Chapter 4. DNA fingerprints are uniquely useful for positive paternity determination because numerous hypervariable loci can be visualized simultaneously as a ladder of DNA bands on a gel, and because the loci follow Mendelian expectations of inheritance over the relevant time scale. This means that all the bands in an offspring's fingerprint should have come from either the mother or the father. So if the mother and offspring are known, the father's pattern of relevant bands can be determined, and paternity assigned with vanishingly small probabilities of misidentification (*Figure 3.3*; and see Chapter 4, Section 6).

Measuring kinship beyond first-order relatives using DNA fingerprints is complicated by a number of factors (27). Most important of these is the loss of information related to the fact that allelic pairs cannot be identified in the multilocus profile. It is preferable to probe for single minisatellite loci, although the procedure for isolating such DNA probes is very labour-intensive (though much less so for the short, highly repetitive 'microsatellite' sequences). From investigations of allozyme polymorphisms, kinship among and within social groups has been measured primarily by two methods: by comparing genetic heterogeneity between groups, and by estimating relatedness among individuals within groups. In the case of the first method, an assumption is made that kin groups should show a non-random distribution of allele frequencies, with greater homogeneity within than between kin groups (see 25). Social groups can be compared at a single locus by the G-test for heterogeneity for evidence of a non-random distribution among groups. Use of the F-statistic allows further partitioning of genetic variance into within-group, among-group, and among-

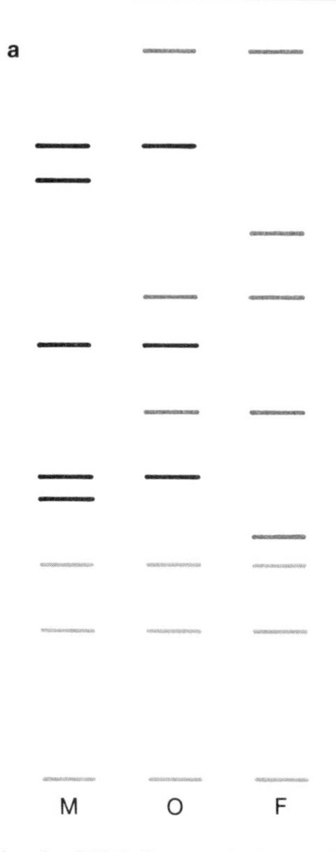

Figure 3.3. Paternity testing by DNA fingerprinting. **a** Schematic representation with paternal bands given in orange and maternal bands in black. **b** Paternity test for red deer with three potential sires (courtesy of Josephine Pemberton). M, mother; F, potential father; O, offspring. The orange arrows indicate 3 paternal bands. F_1 is the father.

population components. A multilocus approach can be applied using a discriminant function analysis as described by Smouse *et al.* (28).

The estimation of relatedness within groups is usually achieved by regression analysis on gene frequencies. Theoretically this can give an average coefficient of relatedness for groups of two or more individuals. The standard error of these estimates is determined by the relative frequency of alleles, the number of individuals in a group and the number of groups used in the regression (29). These estimates of genetic relatedness are based on genetic similarity that could result from either assortive dispersal or common descent. To distinguish these two possibilities, it is necessary to obtain information from natural populations on dispersal patterns.

Appendix 3.1: Gene diversity between populations

For a population divided into s subpopulations, the average gene diversity between subpopulations is given by

$$D_{\mathrm{ST}} = J_{\mathrm{S}} - J_{\mathrm{T}}$$

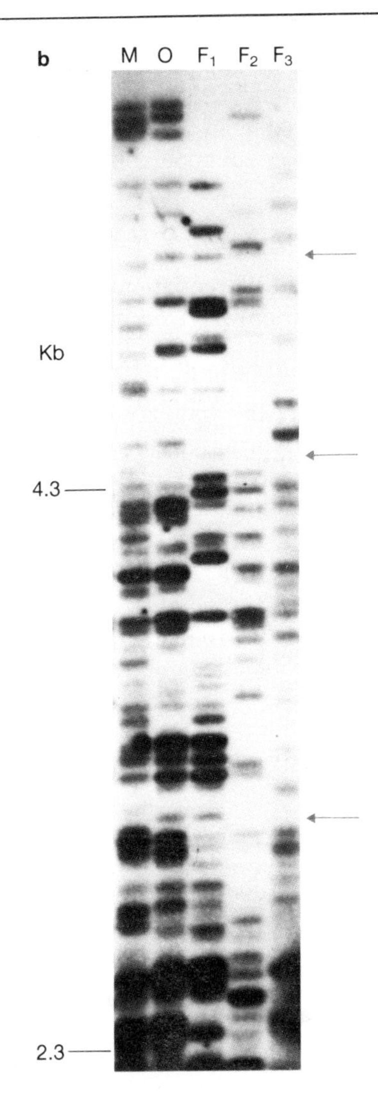

where J_S is the average gene identity (homozygosity) within subpopulations and J_T is the average gene identity for the whole population. Gene identity in a given subpopulation is given by

$$J_i = \sum_k x^2 ik$$

where x_{ik} is the frequency of the kth allele in the ith subpopulation. The gene identity for the total population is

$$J_T = \sum_k x_{.k}^{\ 2}$$

where $x_{.k}$ is equal to $\Sigma_i x_{ik}/s$. The gene diversity (heterozygosity) for the total population is H_T which is equal to $1-J_T$, the average gene diversity for subpopulations is H_S which is equal to $1-J_S$ and $H_T + H_S$ is equal to D_{ST}. Gene diversity between two particular subpopulations (the *i*th and *j*th populations) is given by

$$D_{ij} = ½(J_i + J_j) - J_{ij} = H_{ij} - ½(H_i + H_j).$$

The relative magnitude of gene differentiation among subpopulations (called the coefficient of gene differentiation) is given by

$$G_{ST} = D_{ST}/H_T.$$

However, G_{ST} is dependent on gene diversity. When H_T is very small, G_{ST} may be artificially large. For this reason Nei (8) also describes a measure that is independent of gene diversity, and estimates the minimum net codon differences between populations. This is called the absolute degree of gene differentiation and is given by

$$D_m = sD_{ST}/(s-1).$$

These formulations lend themselves to hierarchical subdivision so that diversity between colonies within subpopulations or demes, etc. can be described (see examples in 8).

Appendix 3.2: Variance of heterozygosity

If the heterozygosity at a given locus in a population is given by *h*, and the number of loci examined is *r*, then the average heterozygosity over all loci is given by

$$H = \sum h_k/r$$

where h_k is the heterozygosity at the *k*th locus. The expected variance of *h* is estimated by

$$V(h) = \sum (h_k - H)^2/(r-1).$$

The sampling variance of *H* is given by

$$V(H) = V(h)/r.$$

This assumes that heterozygosities at different loci are independent, which is generally the case unless there is linkage disequilibrium. Nei and Roychoudbury (7) describe two measures of sampling variance: the interlocus and the intralocus variances. The interlocus variance is determined by diverse evolutionary forces, and is usually very difficult to quantify. The intralocus variance is dependent on the sample size and the gene frequencies of the locus studied. This measure is used to compute the standard errors of heterozygosity and genetic distance, and to estimate the magnitude of interlocus variance. The total variance is equal to the sum of the component variances so that

$$V(h) = V_g(h) + V_s(h)$$

where $V_g(h)$ is the interlocus variance and $V_s(h)$ is the intralocus variance. The intralocus variance can be estimated by

$$V_s(h) = \sum V_s(h_k)/r$$

where r is the number of loci studied. The interlocus variance is estimated by

$$V_g(n-1)^2V_h(h)/n^2$$

where n is the number of genes and V_h is the variance of homozygosity and heterozygosity among loci.

These variance measures are affected by deviation from the Hardy–Weinberg equilibrium, and dominance. Dominance will tend to increase the variance, although the effect is small unless the frequency of recessive genes is very small. Inbreeding is expected to increase the variance as this has the effect of forcing alleles to fixation at random.

8. Further reading

Genetic diversity and distance

Nei,M. (1987) *Molecular Evolutionary Genetics*. Columbia University Press, New York.
Maynard Smith,J. (1989) *Evolutionary Genetics*. Oxford University Press.

9. References

1. Landsteiner,K. (1900) *Zentr. Bakteriol. Parasitenk.*, **27**, 357.
2. Lewontin,R.C. and Hubby,J.L. (1966) *Genetics*, **54**, 595.
3. Harris,H. (1966) *Proc. Royal Soc. Lond.* Ser. B, **164**, 298.
4. Nevo,E. (1978) *Theor. Pop. Biol.*, **13**, 121.
5. Nei,M. (1987) *Molecular Evolutionary Genetics*. Columbia University Press, New York.
6. Nevo,E., Beilles,A., and Ben-Shlomo,R. (1983) *The Evolutionary Significance of Genetic Diversity: Ecological, Demographic and Life History Correlates*. Lecture Notes in Biomathematics, Springer-Verlag, Berlin.
7. Nei,M. and Roychoudhury,A.K. (1974) *Genetics*, **76**, 379.
8. Nei,M. (1975) *Molecular Population Genetics and Evolution*. North-Holland, Amsterdam.
9. Nei,M. (1978) *Genetics*, **89**, 583.
10. Wahlund,S. (1928) *Hereditas*, **11**, 65.
11. Wright,S. (1965) *Evolution*, **19**, 395.
12. Nei,M. (1977) *Ann. Hum. Genet. Lond.*, **41**, 225.
13. Nei,M. and Roychoudhury,A.K. (1982) *Evol. Biol.*, **14**, 1.
14. Nei,M. (1972) *Am. Nat.*, **106**, 283.
15. Upholt,W.B. (1977) *Nucleic Acids Res.*, **4**, 1,257.
16. Nei,M. and Li,W.-H. (1979) *Proc. Nat. Acad. Sci. USA*, **76**, 5,269.
17. Nei,M. and Chakraborty,R. (1976) *J. Mol. Evol.*, **8**, 381.
18. Nei,M. and Tajima,F. (1981) *Genetics*, **97**, 145.
19. Kimura,M. and Ohta,T. (1972) *J. Mol. Evol.*, **2**, 87.
20. Crow,J.F. and Kimura,M. (1970) *An Introduction to Population Genetic Theory*. Harper and Row, New York.
21. Nei,M. and Feldman,M.W. (1972) *Theor. Pop. Biol.*, **3**, 460.
22. Wilson,A.C. and 10 co-authors (1985) *Biol. J. Linn. Soc.*, **26**, 375.

23. Waples,R.S. (1989) *Genetics,* **121**, 379.
24. Hill,W.G. (1981) *Genet. Res.,* **38**, 209.
25. McCracken,G.F. and Bradbury,J.W. (1977) *Science,* **198**, 303.
26. Jeffreys,A.J., Wilson,V., and Thein,S.L. (1985) *Nature,* **314**, 67.
27. Lynch,M. (1988) *Mol. Biol. Evol.,* **5**, 584.
28. Smouse,P.E., Speilman,R.S., and Park,M.H. (1982) *Am. Nat.,* **119**, 445.
29. Pamilo,P. and Crozier,R.H. (1982) *Theor. Pop. Biol.,* **21**, 171.
30. Falconer,D.S. (1981) *Quantitative Genetics*, 2nd edn., Longman, London.

4

Application of molecular techniques to population problems

1. Introduction

In the first three chapters we have briefly reviewed how genetic variation accumulated in a population, how it can be measured in the laboratory, and how measures of variation can be statistically quantified. In this chapter we will describe examples of studies that have employed measures of protein, mtDNA, or nuclear DNA variability to infer kinship and the genetic structuring of populations. It is important to note that interpretations based on measuring variation at each component have some inherent limitations, and further, that non-Mendelian DNA turnover mechanisms can alter patterns of variation in ways that are not predicted by models which assume a constant rate of change over time. For these reasons it is important to compare populations at a number of genetic components, and contrast these results with what is known about the behaviour of DNA at the molecular level, and the behaviour of the organism at the population level.

2. Protein studies

2.1 Introduction

Proteins are composed of one of more polypeptides, which are chains of amino acids. The sequence of amino acids in a polypeptide is determined by triplets of nucleotides (codons) in the gene. The information in the codons is degenerate, so that there are usually several possible codons for a given amino acid. Therefore, a protein may be composed of the product of several gene loci, and fairly substantial changes can take place in the DNA sequence of the gene without changing the composition of the protein. However, small changes in DNA sequence have the potential to alter the structure, and therefore the electrophoretic mobility, of the protein. This has been the utility of investigating protein structure and mobilities as a means of analysing genetic structure: a change in the protein implies a change in the DNA (but see Chapter 2, Section 2). The reverse is not necessarily true. Therefore, the resolution of the technique is not complete. However, the main limitation in comparing populations on the basis of protein studies is the low level of variation in coding sequences relative to other regions of the genome. The advantages are that speculation about function is less tentative than for some of the non-transcribed regions. This is important for investigations on the evolutionary role of natural selection. Further, when protein loci are sufficiently variable to distinguish population structure, the method of protein electrophoresis can offer an inexpensive and relatively expedient means of investigating genetic variation.

Following the development of protein electrophoresis in the mid 1960s (1) the technique was applied extensively to describe levels of variation in a vast array of species (see 2), and to describe the genetic structuring of populations. Methods have been developed to improve both the analytical resolution and the statistical interpretation of protein variation (see Chapter 2, Section 2).

2.2 Isozyme variation in Colias *butterflies: the effect of selection on population patterns*

Two populations of *Colias* butterflies were investigated by Watt and co-workers (3), one in a lowland habitat, the other at 2350 m in the Colorado Rockies. Peak density of butterflies in flight on a given day is related to temperature and wind conditions, and occurs when the butterflies experience a body temperature that is optimal for flight (35 – 39°C). From previous work, Watt (4) had established that certain heterozygotes at the phosphoglucoisomerase (PGI) isozyme locus should be able to sustain flight at suboptimal body temperature, due to greater glycolytic tolerance. Therefore, they predicted that butterflies collected during morning flight at low air temperatures would show greater heterozygosity than butterflies collected the same day at peak flight density.

Butterflies were caught during flight throughout the day on a number of occasions, and as predicted, heterozygotes were more common at low ambient temperature (*Figure 4.1*). Some heterozygous genotypes at PGI were more effective at facilitating flight at low temperature than others. Accordingly, the representation of each genotype varied with mean temperature in catches throughout a given day. Furthermore, the frequency of heterozygotes varied seasonally, and the relative prevalence of frequent and rare alleles varied in different localities with habitat temperature. Taken together these results suggest that the observed pattern is due to natural selection related to heat stress, and not to 'hitchhiking' with an unidentified linked locus, or to other factors that could generate heterozygous excess (see 4).

This example illustrates how important it can be to know the specific ecology of the organism under study when comparing isozyme frequencies at the population level. If a researcher had begun collecting *Colias* butterflies at one location in the morning and finished some distance away in the afternoon, the observed PGI polymorphism would suggest a geographic pattern across the transect, when in reality it represented a behavioural pattern related to the changing temperature over the course of a day.

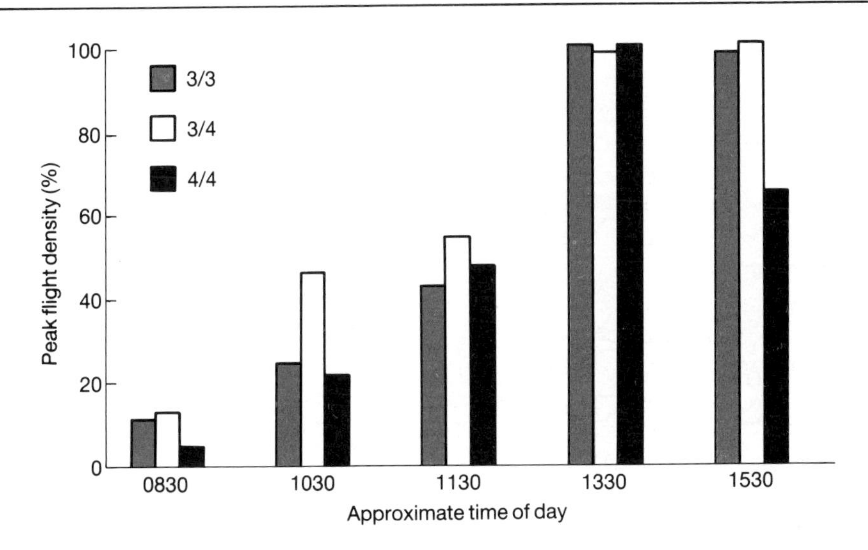

Figure 4.1. Effect of time of day on the representation of three different genotypes (3/3, 3/4 and 4/4) among butterflies in flight (after 3).

2.3 Population structure in brown trout

The brown trout (*Salmo trutta*) is an example of a species that shows considerable genetic diversity at enzyme loci. In some localities, 55 per cent of the total polymorphism is due to differences between populations (G_{ST} is 0.55; see 5). This level of variation has greatly facilitated the description of population structure based on isozyme variants in this species. Karakousis and Triantaphyllidis (6) investigated variation at 25 isozyme loci in brown trout collected from seven streams in northern Greece. Greek trout had been classified into five different subspecies based on minor differences in morphology. The genetic data indicated that only one of the seven populations differed substantially from the other populations. Most comparisons between rivers and between putative subspecies gave genetic distance values comparable to those reported for comparisons between other populations of the same species, and were not consistent with the level of differentiation expected at the subspecific level. The one exception had a genetic distance 3–5 times greater than for other pairwise comparisons between populations, and a unique allele at the creatine kinase locus. These fish (*S.t.peristericus*) came from an isolated stream which flows to a lake in the north of Greece. All of the other populations inhabited rivers that flowed (directly or indirectly) into the Mediterranean (*Figure 4.2*).

On a broader scale, Karakousis and Triantaphyllidis (6) compared published data for isozyme variation in 19 populations of European brown trout. They identified two principal groups separated by a genetic distance of about 0.14, one in northern Europe (Britain, France, and Sweden) and the other in the Mediterranean and the Black Sea. They suggest that this supports an earlier hypothesis (7) that the brown trout from the Mediterranean basin descended

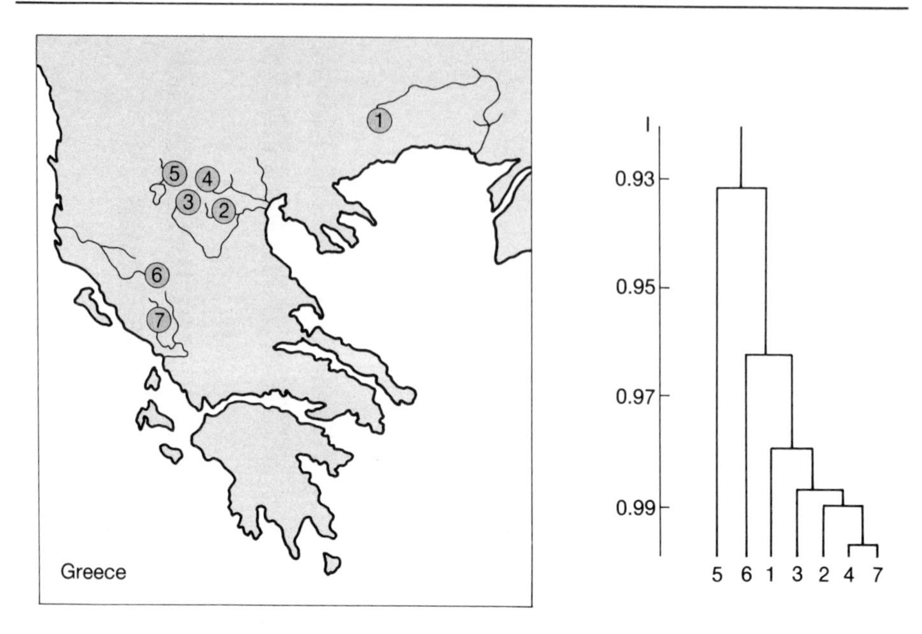

Figure 4.2. Location of study populations of brown trout in Greece, and the level of genetic identity between them as indicated by allozyme variation (after 6). Not all rivers are shown.

from a hypothetical anadromous subspecies that disappeared from the sea about 300 000 years ago, during the last interglacial period.

3. Mitochondrial DNA

3.1 Introduction

In multicellular animals the mitochondrial genome is a circular, double-stranded molecule ranging in size from 15.7 kb to 19.5 kb. It is functionally different from the nuclear genome in a number of respects. For example, replication is asymmetric, unidirectional, and continuous, requiring far fewer enzymes than the symmetric, bidirectional, discontinuous replication of nuclear DNA (8). The gene content is apparently invariant across all metazoans studied so far (primarily vertebrates and *Drosophila* spp.) and limited to 13 proteins, 2 ribosomal RNAs, and 22 transfer RNAs. Replication and transcription are initiated in the 'control region' where variable non-transcribed regions are also concentrated.

It has been demonstrated that mtDNA is inherited maternally, by transmission through the egg cytoplasm (see 9). Present evidence suggests that there is effectively strict maternal inheritance without 'paternal leakage', although it is still an open question. Maternal inheritance was determined through cross-breeding experiments where the maternal and paternal mitochondrial genomes differed. For example, horses and donkeys have different mtDNA *Hae*III restriction patterns. A cross between a female horse and a male donkey produces a mule with horse mtDNA. The reciprocal cross produces a hinnie with donkey mtDNA (10).

The questions of maternal inheritance and apparent haploidy of mtDNA are of central importance. There are about 10^5 mitochondria in a mammalian egg, and about 50 in the midpiece of the sperm. If the sperm contributes no mitochondria to the subsequent generation, and the mtDNA in the egg is homogeneous, then mtDNA will be transmitted as a haploid genome, and only within matrilines. This would make mtDNA a powerful genetic marker for population studies. This interpretation depends on complete homoplasmy (no intraindividual variation in mtDNA). Theoretically heteroplasmy could arise either through mutation or by paternal contribution. There is some indication that this may occur rarely in some vertebrate species. This could have important consequences for the interpretation of genealogies (see discussion in 11).

For species where there are sex-biased dispersal patterns such that females tend to be philopatric, a comparison with variation in nuclear DNA would show much greater differentiation in the maternally transmitted genome (mtDNA). This could be employed to help resolve investigations where sex-biased dispersal was suspected.

3.2 mtDNA variation

Sequence changes in animal mitochondrial genomes are of four principle types: sequence rearrangements, additions, deletions, and nucleotide substitutions. Overall substitution rates for the mitochondrial genome have been estimated to be 5–10 times greater than in 'single-copy' nuclear DNA (12). The lowest mtDNA substitution rates are in the tRNA and rRNA genes. The mtDNA protein genes evolve at about twice that rate (which can be up to two orders of magnitude higher than their nuclear counterparts). Rates vary considerably between proteins and at a given protein among different species. It has been suggested that the mean rate of divergence for the mitochondrial genome over a wide range of taxa is 2 per cent per million years (11). This estimate was derived from studies where evidence on species divergence (e.g. from fossils) was already available. However, since rates vary, the rate for a particular region or species should be established before interpreting distance measures at the population level. Further, many bases are conserved in the mitochondrial genome, so the substitution rate will level out as divergence time increases and the more variable sites become saturated (12; *Figure 4.3*).

There are non-coding sequences in the mitochondrial genome, although proportionally far fewer than in the nuclear genome. Most of these sequences are found immediately adjacent to structural genes and are quite small (usually 5 bp or less in vertebrates). Substitution at these sites occurs at about the same frequency as for synonymous third position codon sites in protein genes (which evolve at 3–4 times the rate of non-synonymous codon sites; see 13).

The most variable part of the mitochondrial genome for additions and deletions is the region where replication begins, known as the control or D-loop region.

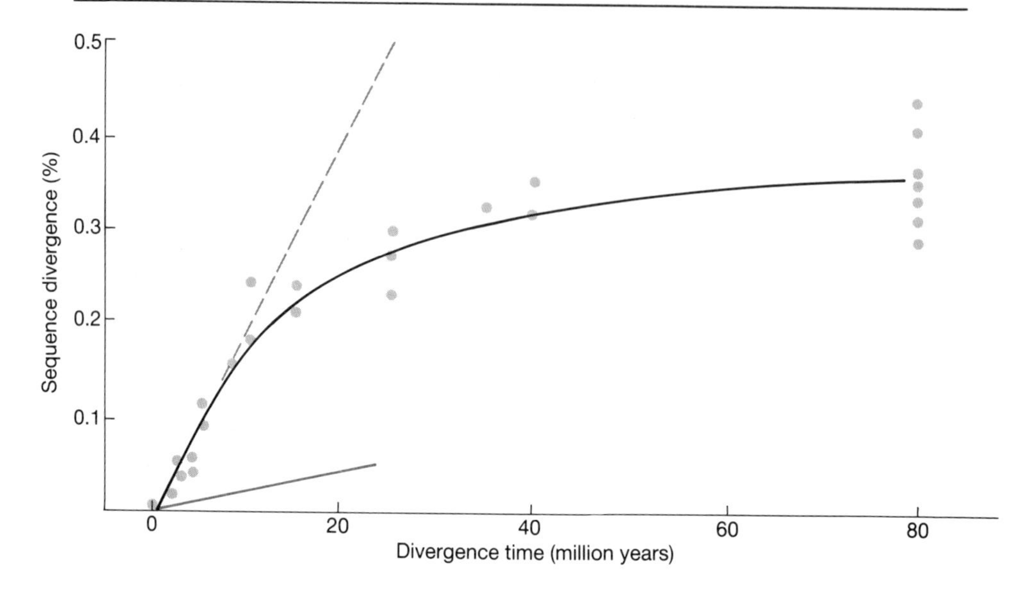

Figure 4.3. Rate of nucleotide substitution in mtDNA (black lines) and single-copy nuclear DNA (orange line) (after 12).

It is estimated that the size of this region varies among animal species from about 200 to 4100 bp (8). There is evidence that the process of slippage is operating in portions of the primate and cetacean control regions (14). The substitution rate in the human D-loop region is estimated to be 2.8–5 times the rate found in the remainder of the mitochondrial genome, though this rate disparity is not seen for interspecific comparisons in primates, rodents, and cetaceans (see 14). There are several conserved blocks near the promoter sequences, one of which has been associated with a function related to heavy-strand replication (15).

Bottlenecks can greatly influence the level of mtDNA variability. For example, if a population of diploid animals is reduced to a single breeding pair, they will have four copies of the nuclear genome, but only one transmissable copy of the mitochondrial genome. Assuming homoplasmy for mtDNA and no paternal leakage, variation in mtDNA will be eliminated, while for brief bottlenecks significant nuclear variability can be retained. This could have a dramatic affect on the interpretation of genetic distance between populations. If a rare genotype was fixed by a founder event in one of the populations being compared, the apparent distance would indicate far greater genetic division than was justified. A number of species show low levels of mtDNA variation compared to nuclear DNA, suggesting the possibility of a bottleneck period. For example, the anomalously low level of variation in human mtDNA (see Section 3.4) has led to the speculation that a transient bottleneck was involved in the formation of *Homo sapiens* (16).

3.3 The cichlid fishes of Lake Victoria

Lake Victoria and its smaller satellite lakes in Uganda, Tanzania, and Kenya are less than 1 million years old, yet they are inhabited by more than 200 endemic cichlid fish species. The morphological variation between these species is not extreme, but there are a great variety of ecological specializations. These include some truly bizarre adaptations, such as one species that plucks the eyes from other fish, or another that feeds on the scales from the fins of other fish. Some species raise their brood in the mouth, and others specialize on engulfing the mouth of the mouth-brooders and sucking out the brood. Some biologists have interpreted this as an example of an explosive radiation of new species, while others speculate that it represents a polyphyletic invasion of taxa that had become differentiated before the lake's formation.

Meyer and co-workers (17) sequenced up to 803 bp of mtDNA from 14 representative species in Lake Victoria and 23 additional African species. They used the PCR technique (see Chapter 2, Section 6) to amplify a 363 bp fragment from the cytochrome *b* gene, and 440 bp from a region including tRNA genes and sequence from the non-coding D-loop region. The species selected from Lake Victoria included representatives of each of the main ecological and morphological groups, namely eaters of fish, fish larvae, molluscs, insects, and algae.

For the 14 Lake Victoria species, only 15 out of the 803 bp surveyed showed any variation, ranging from 0 – 5 bp differences between species. This represents a genetic distance of less than 0.5 per cent for most comparisons. However, there was little sharing of mtDNA types between species, a very unlikely result unless these truly represent different species. The differences between species in different lakes were far greater, and consistent with the estimated geologic age of the lakes. Assuming a rate of change of 2.5 per cent per million years, Meyer and co-workers estimate that the Lake Victoria species originated about 200 000 years ago, and have diverged from cichlids in Lake Malawi by 1 – 2 million years, and from cichlids in Lake Tanganyika by 2 – 4 million years. The age of Lake Victoria has been estimated at anywhere from 250 000 to 750 000 years. These results support the idea that the African lake cichlids represent recent radiations of morphological diversification at a remarkable rate, and without an acceleration of molecular evolution (at least of the sequences under comparison).

3.4 Human evolution

Cann and co-workers (18) isolated whole genomic mitochondrial DNA from 147 people representing five geographic populations. There were 20 Africans, 34 Asians, 46 Caucasians (originating from Europe, North Africa, and the Middle East), 21 aboriginal Australians, and 26 aboriginal New Guineans. They compared RFLP patterns (see Chapter 2, Section 3) using 12 restriction enzymes that have 4- or 5-bp recognition sites. This allowed the description of a phylogeny based on 195 polymorphic sites, out of a total of 467 sites identified. An average individual was investigated at 370 sites, which represents about 9 per cent of the 16 500 bp human mitochondrial genome.

They used a parsimony method to build a tree relating 133 restriction pattern types, and present one of a number of possible trees. They infer from this tree that Africa is the likely source of the human mitochondrial gene pool. This follows from the observation that both primary branches of the tree lead ultimately (and in one case exclusively) to African mtDNA types (*Figure 4.4*), though there are relatively few changes distinguishing the primary African clade from the rest of the tree. A second implication from the observed patterns is that each of the non-African populations has multiple origins. This was determined by counting

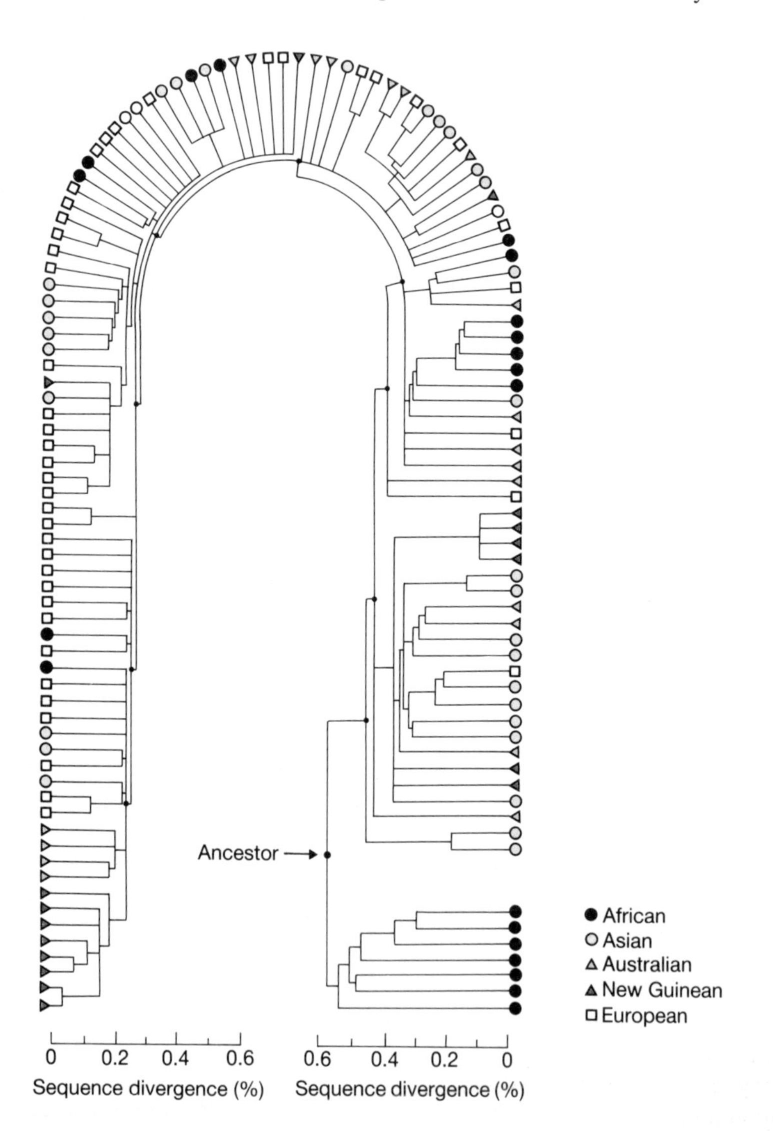

Figure 4.4. Genealogical tree for 134 types of human mtDNA from RFLP data (after 18).

the number of region-specific clusters for each group, which ranged from 15 to 36. The mean pairwise genetic distance within clusters ranged from 0.09 in Europe to 0.36 in Africa. These distances are fairly low, and have suggested to some (16) that there may have been a transient or prolonged bottleneck in human evolutionary history. If the human mitochondrial genome is evolving at 2–4 per cent per million years, as has been found for other mammalian taxa, then the clusters can be aged. This would suggest that the European cluster was founded as recently as 23 000–45 000 years ago, and that the migrations from Africa could have been as much as 90 000–180 000 years ago.

4. Nuclear DNA: single-copy genes

4.1 Introduction

Whereas most of the mitochondrial DNA of animals can be assumed to be changing in a more or less regular fashion by point mutations, and can thus be exploited for tracing animal lineages through real time as in the above examples, the analysis of DNA variation in nuclear genomes can be much more complicated. This is because most animal and plant nuclear genomes contain vast amounts of additional DNA over and above the coding requirements of the organism. They contain thousands of repetitive families which differ widely in copy-number, length of repeat, sequence composition, and dispersion around the chromosomes. Further, nuclear DNA is affected on an evolutionary time-scale by a variety of DNA turnover mechanisms, as described in Chapter 1, Section 5.3. These increase the number of different types of mutation that can occur, and can promote or demote the representation of variant sequences in a population. Therefore, the analysis and interpretation of genetic variation in nuclear genomes requires both an understanding of the internal dynamics of DNA mutation and turnover, and an understanding of the external processes of selection and drift at the population level.

It has been generally assumed that the single-copy state of a gene lessens the need to consider DNA turnover when interpreting intraspecific polymorphism and interspecific divergence. However, the analysis of, for example, human globin gene and HLA polymorphisms has revealed the activities of small domains of gene conversion and slippage (see Chapter 1, Section 5.3) which play a significant role in the levels and patterns of mutant distribution in these two genetic systems. Therefore, these types of patterns should be identified and quantified to allow for the accurate interpretation of genetic differences at the population level.

4.2 Alcohol dehydrogenase variation in Drosophila

One of the most intensely studied genes in animal species has been the *Adh* locus coding for alcohol dehydrogenase. Proponents of natural selection and neutral theory have both based arguments on the analysis of this gene. *Adh* is a single-copy gene (i.e. it has two alleles in the diploid phase), although some species of *Drosophila* have duplicate copies. The *Adh* locus in *D.melanogaster* has four exons (coding regions) and three small introns (non-coding regions) (*Figure 4.5*).

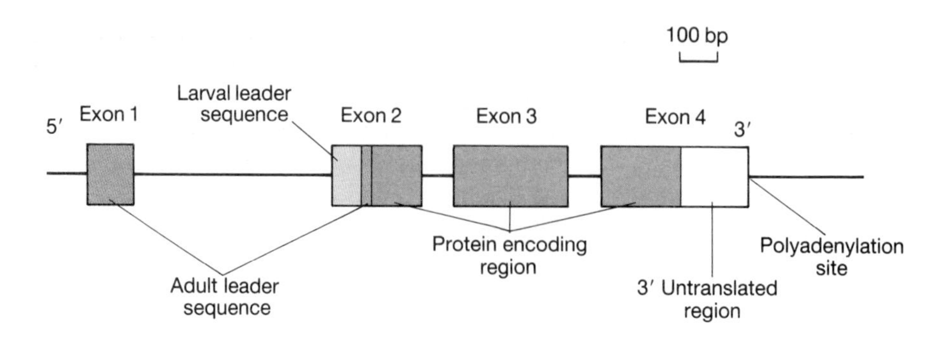

Figure 4.5. Structure of the *Adh* gene of *Drosophila melanogaster* (after 20).

Kreitman and Aguadé (19,20) investigated variation at the *Adh* locus by cutting with 4-bp recognition site enzymes, running the samples on a polyacrylamide denaturing gel, and probing with the cloned *Adh* locus (see Chapter 2). They ran comparisons with 10 different enzymes. This allows discrimination based on fragment size at a fine level of resolution. Further, running samples side by side allows the detection of deletions and insertions. The positions of restriction sites were determined by comparing the RFLP patterns with the known sequence of the region.

They compared 87 isofemales lines (a homogeneous group of individuals all derived from one female) of *D.melanogaster* representing two natural populations (the east and west coasts of North America). This revealed 50 distinct haplotypes, but no evidence for genetic differentiation between the two populations. In a separate study, Kreitman (21) sequenced 11 cloned *Adh* genes from five separate natural *D.melanogaster* populations and found 43 polymorphisms, only one of which resulted in an amino acid change. Closely related alleles were distributed over a broad geographic range, suggesting that the effective population size (N_e; see Chapter 3) is large. Based on this sequence data, Kreitman estimates N_e for *D.melanogaster* to be 3×10^6, which is very large. This is important, because when N_e is close to the estimated species population size (as in this case; see 21), then selection will be more important than drift in governing the evolutionary fate of slightly deleterious mutations. A slightly deleterious mutation that may become fixed by chance in a smaller population would be more likely to be eliminated by selection in a large panmictic population. The relative importance of genetic drift increases dramatically when the effective population size becomes small (see Chapter 3).

D.melanogaster has been found to have a polymorphism for two isozyme alleles in nearly all natural populations that have been studied: a 'slow' and a 'fast' allele which differ by a single amino acid. In Kreitman's study, the DNA sequence of five fast and six slow alleles were compared. There were many silent polymorphisms (in third-position codon sites that do not alter the amino acid), and changes in introns, but no changes that resulted in the replacement of an amino acid, other than the one that represents the electrophoretic migration difference between the two isozyme alleles. This result suggests strong selection for the conservation of the amino acid sequence. It also emphasizes the wealth of 'hidden' variation not detected by isozyme analysis. The extreme difference between the low rate of change in replacement sites (leading to a change in the amino acid) and the high rate in silent sites could be due in part to the large N_e of this species, as described above.

5. Nuclear DNA: gene families

5.1 Introduction

One of the most useful components of nuclear genomes for the identification of individuals, populations and higher units is the multigene family coding for the 28S and 18S ribosomal RNAs. The family consists of a repetitive unit containing one copy of each of the two major genes separated by an intergenic spacer. The compound unit of genes and spacer are usually repeated from several hundred to thousands of times in tandem arrays that can be on several pairs of chromosomes, depending on species. For example, there are five pairs of rDNA arrays in humans.

The utility of the rDNA gene-family for the study of natural variation, population structure, and breeding behaviour derives from two features. First, each spacer is often further divided into a tandem array of subrepeats whose precise lengths differ between species. Some spacers, such as those in *Drosophila* and *Xenopus*, contain several different arrays of subrepetition (*Figure 4.6*).

Second, unequal crossing-over (see *Figure 1.3*) is known to be occurring both within the spacer repeats and between whole units in the array. Unequal crossing-over at the periodicity of a unit of subrepetition in the spacer leads to the continual gain and loss of numbers of subrepeats, which can be detected as different spacer lengths by cutting the DNA at restriction enzyme sites that lie outside the array of subrepeats. Unequal crossing-over at the periodicity of the whole rDNA unit leads to variation in the copy-number of a particular length generated at the lower level. This can be detected by the intensity of bands in a gel restriction digest probed for the rDNA spacer. Hence, the position of a band is indicative of the length of the fragment (which is a reflection of the number of spacer subrepeats); and the intensity of a band is indicative of the number of whole units of a given length. The absence of length variation in any given population is probably a consequence of the absence of IGS subrepeat arrays, rather than the result of population bottlenecks and inbreeding, as has been occasionally suggested. The two alternative interpretations are easily tested experimentally.

Measured rates of unequal crossing-over both within and between different chromosomal arrays are approximately 10^{-4} per generation per rDNA unit in *Drosophila* (22) and yeast (23). These rates are faster than the base substitution rate (and hence lead to the homogenization of mutations through the arrays and species), but are too slow to generate new spacer lengths at a rate which would disturb their use for the identification of parent-offspring.

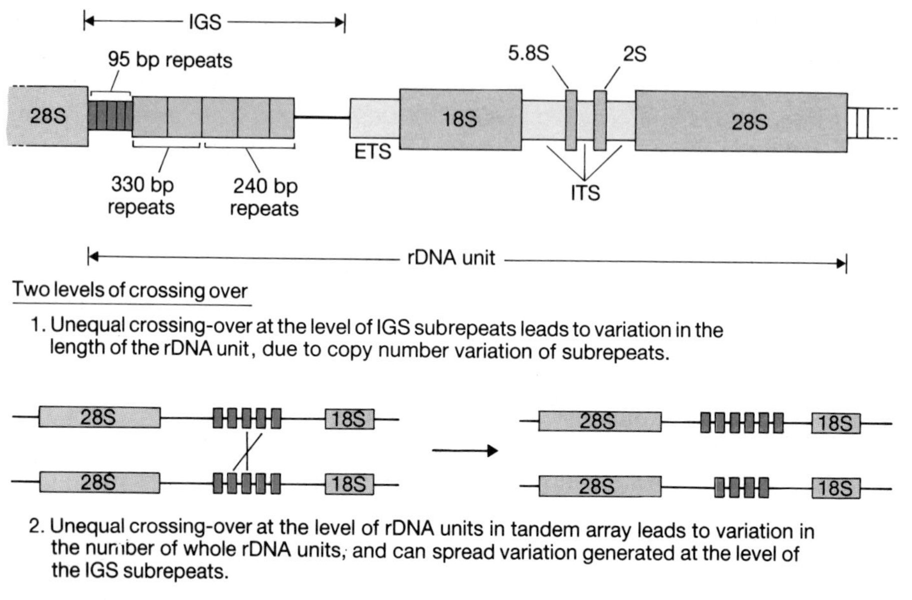

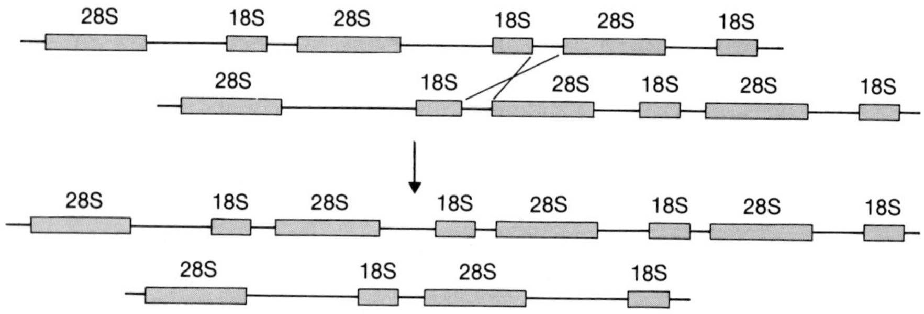

Figure 4.6. A typical rDNA unit from *Drosophila melanogaster*, and the generation and homogenization of variation by unequal crossing-over. IGS, intergenic spacer; ITS, internal transcribed spacer, ETS, external transcribed spacer (after 22).

In a study of rDNA spacer variation amongst species of *Drosophila*, Coen and colleagues showed that the 500 copies per individual were homogeneous for mutations that were diagnostic for a species (22). For example, each of the spacer subrepeats in each of the spacers in *D.melanogaster* contained a restriction enzyme site not present in *D.simulans* and several other related sibling species. In humans and mice it has been shown that the rDNA families are divisible into subfamilies on the basis of diagnostic restriction sites (24). These might represent partially homogenized mutations or they might reflect the restriction of unequal crossing over to a subset of the available repeats. Chromosomes, for example, might be a natural barrier to unequal crossing-over in species where the rDNA array is divided amongst several non-homologous pairs of chromosomes. In humans, some of the subfamilies are evenly distributed amongst the five chromosomal locations indicating that, in these cases, the chromosome is not a barrier to exchanges by unequal crossing-over, or the possibility of gene conversion. There are other partially homogenized mutations, however, that are restricted to particular rDNA arrays in human populations. In mice the rDNA subfamilies are chromosome-specific, although mouse satellite DNA has subfamilies shared amongst all chromosomes (25). Such studies indicate that the evolutionary history and subsequent distribution of genetic variation from chromosomes upwards, is different for each family in each species; no generalizations can be made. Furthermore, it is important to take into account the age of a mutation as well as its distribution, before drawing conclusions on possible restrictions to interchromosome homogenization (see 26). For example, the chromosome-linked rDNA variants in humans might be recent relative to other mutations that have had time to spread around the rDNA arrays.

Despite the extensive homogenization (or partial homogenization) of mutations in rDNA families and the corresponding reduction of variation within but not between species, there is a great deal of variation to be exposed due to the gain and loss of copy-numbers of spacer subrepeats and whole rDNA units, as described above. This variation can be detected with the use of appropriate restriction enzymes on whole DNA and by probing the resultant gels with different regions of the rDNA unit by the Southern hybridization technique (see Chapter 2).

5.2 rDNA variation in wheat

Over the past few years rDNA has been extensively used as a genetic marker in diverse species, especially in plants. For example, Flavell and co-workers (27) have examined the distribution of spacer length variants in 112 plants taken from 12 populations of wild wheat (*Triticum dicoccoides*) for which allozymic variation encoded by 50 gene loci had been previously established. Populations of wild wheat are geographically structured, and the distribution of allozymes suggested that selection was responsible for some of the differences in localities with different climates and soil types (28).

The rDNA family in wheat is distributed on two pairs of chromosomes, and spacer length variation is due to the gain and loss of a 135-bp subrepetition within the spacer. The results of the survey show that natural populations of *T.dicoccoides* display a wide spectrum of spacer lengths, with some populations being very homogeneous (that is, all arrays of rDNA on the two non-homologous chromosomes have the same spacer length); while other populations have either intermediate levels or very high levels of heterogeneity. The allozyme and rDNA diversities are significantly intercorrelated both between themselves and with the climatic variables.

The highly heterogeneous populations had nine or more different lengths of spacers, whereas the homogeneous populations displayed a single length that was the most frequent length in the heterogeneous populations. These results emerged with the use of a single restriction enzyme *Taq*I. The further use of *Eco*RI + *Bam*HI, *Dde*I, and *Hin*fI showed that there are at least three major types of rDNA repeat in the homogeneous population. This illustrates how the use of one restriction enzyme may underestimate the heterogeneity within the individual. However, the use of *Taq*I has not underestimated the heterogeneity within the population, since, with all enzymes used, all individuals within a homogeneous population were identical. The additional enzymes were revealing additions and deletions of DNA within the spacers, but outside the array of 135-bp subrepeats.

The average number of electrophoretic alleles per population, the proportion of polymorphic loci per population, and levels of genetic diversity are highly correlated with rDNA variables, using Pearson correlations, in particular between the genetic diversity index and the number of independently occurring spacer lengths. A full explanation of the forces responsible for these distributions is not possible from the data available so far. However, *T.dicoccoides* displays a highly subdivided population structure with only limited gene flow between semi-isolated populations, which can be characterized by peaks of locally common alleles. Populations with high allelic variability and rDNA variants are situated in climatically highly fluctuating regions.

Is selection acting on the rDNA length variants directly or on alleles tightly linked to them? Species of *Drosophila, Xenopus,* and wheat are known to contain functionally important signals for transcription within the spacer subrepeats, the copy-number of which affects transcription levels (see 29). It could be that this is the functional basis on which selection can act. However, it is unlikely that selection can act on the first variant spacer occurring in a family of several hundred members. Some appreciable level of homogenization and fixation would need to take place, as a consequence of unequal crossing-over in this case, before appropriate effects on phenotype are 'visible'. Some means of generating the same spacer length variant on different chromosomes must also be operating. From what is currently known in *D.melanogaster* this, too, could be inter-chromosomal unequal crossing-over and/or gene conversion.

6. Nuclear DNA: hypervariable minisatellites and DNA fingerprinting

6.1 Introduction

The analysis of breeding systems is greatly facilitated by utilizing DNA fingerprints to establish pedigrees (see Chapter 3, Section 7), and this is an important component to understanding and predicting dispersal patterns and other mechanisms of population mixing.

Minisatellites derive their name from their existence as relatively short tandem arrays of repeats scattered on all chromosomes. Hypervariability is detectable as variation in copy-number of repeats at the different loci. This is analogous to the high variability in copy-number of subrepeats in each rDNA unit which gives rise to spacer length variation (see Section 5). As with the rDNA, differences between individuals with respect to the repeat copy-number in each

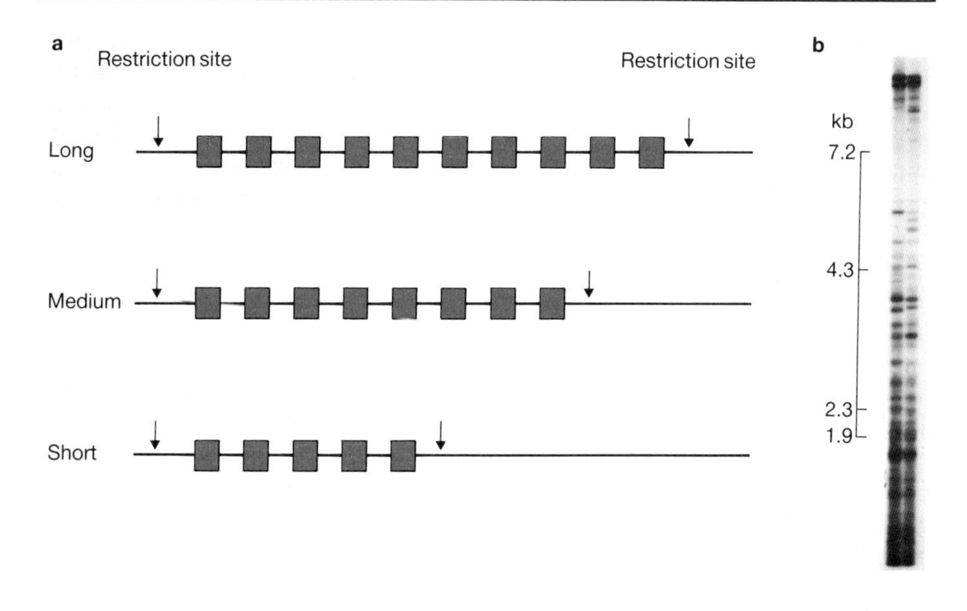

Figure 4.7. Schematic representation of the variation in the lengths of minisatellite arrays **a**, which leads to the variability of DNA fingerprints **b**. DNA fingerprints shown are from two killer whales (after 37).

array can be detected by the use of a restriction enzyme that cuts outside the array and not within any of the repeats (*Figure 4.7*). Hence, variation in the distance between the two given restriction sites gives rise to variable DNA fragments in a gel which can be detected after hybridization to a probe of the repetitive unit.

Jeffreys and co-workers (30) were the first to describe the technique of simultaneously probing for numerous hypervariable minisatellite loci. The initial minisatellite they isolated was comprised of four tandem repeats of a 33-bp sequence in the intron of the human myoglobin gene. A probe was prepared by the purification of a single 33-bp repeat element, followed by the construction of a 'polymer' consisting of 23 of these repeats ligated together. Using this polymer as a probe against *Hin*fI or *Hae*III restricted total DNA, multiple DNA fragments were detected as well as the parent DNA fragment from the myoglobin gene. Differences in the size of fragments both within and between individuals are a result of copy-number variation of the repeats at different loci, because neither *Hin*fI nor *Hae*III cut within the repeats themselves. Cloning and isolation of the minisatellite from different loci showed that repeats from different loci are related by a shared central 'core' sequence.

The discovery of DNA 'fingerprinting' by Jeffreys and co-workers has led to the identification of other hypervariable minisatellites using a range of probes. These have included naturally occurring probes isolated from specific cloned regions of a genome; the use of bacteriophage M13, and simple sequence DNA (microsatellites).

DNA fingerprint fragments are, to all intents and purposes, stably inherited in a Mendelian manner. This holds true so long as the rate of production of new length variants is as low as $0.5-1.5\times10^{-4}$ per gamete per kb of minisatellite. At this rate, it is highly improbable that the DNA fingerprints of a few related individuals in a defined pedigree over a few generations will reveal the accumulation of novel fragments via the non-Mendelian transmission consequences of unequal crossing-over and/or slippage. After long periods of time, however, the multiple length variants produced by either mechanism would have accumulated in the population, which is why the minisatellites are currently observed to be hypervariable. It might not be justified, therefore, to employ methods of analysis of population variation based on Hardy – Weinberg equilibria to minisatellite variation, because the frequencies of variants in the population are probably determined by both the Mendelian transmission of chromosomes and the non-Mendelian behaviour of the DNA. Further, too much variability can be detrimental for race and population identification, because the high rates of generation of new variants in each population will tend to obliterate any population-specific variants that might have differentiated the populations at the time of inception.

For the purposes of linking minisatellite loci to other genetic markers, and improving the accuracy of kinship assessment (see Chapter 3, Section 7), use can be made of clones of locus-specific arrays of minisatellites (which resolve just two bands per individual). One particular array has proven useful in this respect for human forensic studies, because it is an extremely polymorphic locus (heterozygosity = 97 per cent) isolated from a single band (restriction fragment) in a human DNA fingerprint (31). The locus shows extreme length variation due to allelic variation in the number and slight differences in the length of the repeat unit. In a random sample of 158 chromosomes, one common and 76 rare alleles could be resolved. The estimated rate of production of new length variants (assuming N_e for humans to be approximately 10^4) is 0.002 per gamete. The average length of minisatellite DNA at the locus is 5 kb, and thus the rate per kb is 4×10^{-4}; this compares to rates of 10^{-4} from other loci (see above).

Jeffreys and co-workers (32) have also devised a method for investigating variation within a given minisatellite array. That is, they look for variation in the distribution pattern of a variant repeat (identified by a restriction enzyme site) in the array, and compare this pattern between individuals. This has revealed another extensive source of measurable variation at these loci. Further, by using the PCR technique (see Chapter 2, Section 6) to amplify DNA from a number of large deletion mutants in individual sperm from a single human male, they showed that exchanges between alleles at meiosis did not occur in the generation of those mutants. They suggest that alleles at this locus evolve independently on separate chromosomes by mechanisms such as slippage and unequal sister chromatid exchanges. If this were strictly the case then such alleles would be powerful markers for monitoring the fate of haploid chromosome lineages in a population and assessing post-demographic movements. Before this could be achieved it is necessary to check, using appropriate analytical techniques, that sampled chromosomes do not share a particular mutant distribution pattern as a consequence of interallelic gene conversion.

6.2 Dunnock mating behaviour and parental care

Davies and co-workers (33) have studied the breeding behaviour of a population of about 80 dunnocks (*Prunella modularis*) in the Cambridge botanical gardens. Perhaps unexpectedly for such a drab and inconspicuous bird, the dunnock has one of the most complex of all known avian breeding systems. Females hold territories which may be defended by one male (monogamy), by two unrelated males (polyandry), or sometimes by two males jointly defending two or three contiguous female territories (polygynandry). When two males jointly defend a female, there is a clear hierarchy, and the alpha male is able to displace the beta male. Monogamous males guard their mates closely, and effectively exclude most potential competitors, but when two males share a female competition for matings is intense.

Burke and co-workers (34) investigated this same population, applying the technique of DNA fingerprinting (see Chapter 3, Section 7) to determine paternity and thereby the relative reproductive success of different males. They used the multilocus probe 33.15 (30) to test for paternity at 15 monogamous nests (49 chicks), 11 polyandrous nests (34 chicks), and 19 polygynandrous nests (50 chicks). There were an average of 23 bands per fingerprint, an average probability of bandsharing, x, of 0.24, and 5–18 paternally derived bands per paternity test. The probability that an unrelated male would contain all of the paternal bands is $P = x^y$, where y is the number of paternal bands. In this case P ranged from 8×10^{-4} ($y = 5$) to 7×10^{-12} ($y = 18$). For a close relative (a brother, father, or son) $P = [\frac{1}{2}(1+x)]^y$, and in this case ranged from 0.092 to 1.8×10^{-4}.

Out of a total of 133 paternity tests there was only one instance of a male, other than one defending at a nest, achieving a mating. This is in contrast to the high degree of extra-pair fertilization uncovered by DNA fingerprinting in other passerine species (35). There were three main findings from the analysis of paternity at polyanderous and polygynandrous nests. First, both the alpha and beta males achieved matings; second, the proportion of time a male had access to the female was a good predictor of his chances at paternity; and finally, males were more likely to assist at feeding the chicks if they had paternity. There was no indication that males could recognize their own offspring, rather they apparently used mating access as an indicator of how much effort they should invest in feeding the chicks. Davies and Houston (36) have calculated that an alpha male would need to achieve 60–70 per cent of the matings for cooperative polyandry to give him greater reproductive success than monogamy. Burke and co-workers (34) show that for 20 broods where both males had access to females during the mating period, the alpha male achieved 55 per cent of the matings, less than the proposed critical level. They conclude that females benefit because chicks are fed more, but males would be better off with an exclusive association. Consistently, they observe that females encourage the presence of beta males, while the alpha males try to chase them away.

6.3 Inbreeding and population structure

The California Channel Islands lie 25 – 100 km off the coast of southern California near Los Angeles. There is a dwarf fox species (*Urocyon littoralis*) that lives only on these islands. Gilbert and co-workers (37) used the multilocus minisatellite probe 33.6 (30) to compare individual foxes within and between six island populations. They found that the average percentage difference (APD) of fingerprint profiles compared between individuals on the same island was much lower than for comparisons between islands (0.0 – 25.3 per cent versus 43.8 – 84.4 per cent; APD is the average of all *D* values, where *D* is the number of fragments that are different between two individuals divided by their combined total number of fragments, multiplied by 100). The smallest islands had the lowest intra-island APD values, and the islands that were closest together had the lowest inter-island APD values. On some islands there were restriction fragments unique to that island and shared by all individuals. These results suggest inbreeding in the isolated island populations. Gilbert and co-workers construct a phylogeny and suggest that the resulting tree is consistent with the archaeozoological and geological record relevant to the colonization of these islands by mainland foxes, about 17 000 years ago (*Figure 4.8*).

A similar pattern of minisatellite bandsharing was seen for a comparison of five putative killer whale (*Orcinus orca*) populations (38). In this case, a long-term observational study of individually known whales had established that there were potentially three populations in the Eastern North Pacific near Vancouver Island, British Columbia. Two of these populations preyed on fish, but were separated by a tidal boundary at a river estuary. None of the individuals in these two populations had been seen to interact. The third population preyed on marine mammals and was sympatric to each of the other two. These three possible

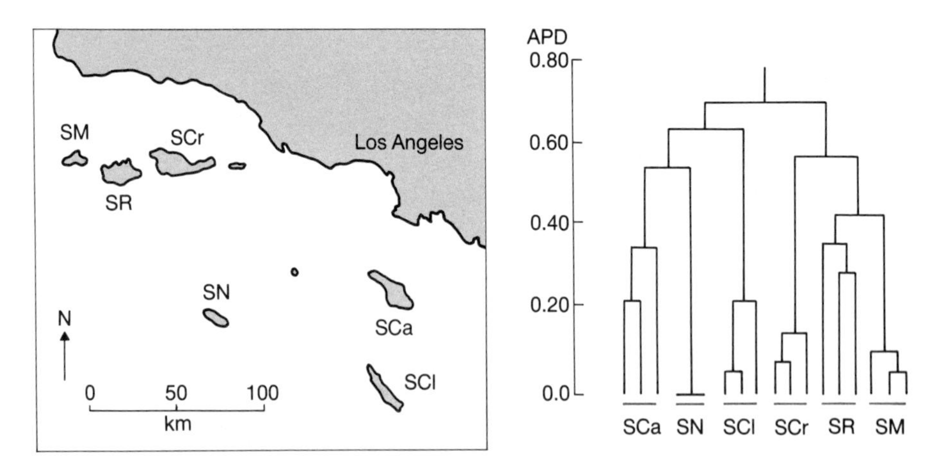

Figure 4.8. Phylogeny for Channel Island foxes based on the average per cent difference (APD) between DNA fingerprint patterns (after 36). Compared between six islands: San Miguel (SM), Santa Rosa (SR), Santa Cruz (SCr), San Nicolas (SN), San Clemente (SCl), and Santa Catalina (SCa).

populations were compared with killer whales from Iceland and Argentina. Minisatellite profiles revealed by hybridizing with the DNA probe 33.15 (30) showed APD values within populations ranging from 23 per cent to 45 per cent. Most between-population APD values were in the range 66 – 73 per cent. The one exception was the comparison between the two Eastern North Pacific populations that prey on fish on either side of a tidal boundary: their APD was 28 per cent. The distinction between the sympatric populations (APD = 67 – 71 per cent) was also seen in a comparison of mtDNA variation, which implied a genetic distance comparable to that seen between allopatric populations (e.g. Eastern North Pacific vs. Western South Atlantic). These results were consistent with the observations that local killer whale populations often show distinct morphological variation, and that at least in the Eastern North Pacific, the rate of dispersal from local 'communities' is either non-existent or very low.

7. Further reading

Dover,G.A. and Flavell,R.B. (ed.) (1982) *Genome Evolution.* Academic Press, London.
Nei,M. and Koehn,R.K. (ed.) (1983) *Evolution of Genes and Proteins.* Sinauer, Sunderland, MA.
MacIntyre,R.J. (ed.) (1985) *Molecular Evolutionary Genetics.* Plenum Press, NY.
Nei,M. (1987) *Molecular Evolutionary Genetics.* Columbia University Press, NY.

8. References

1. Lewontin,R.C. and Hubby,J.L. (1966) *Genetics,* **54**, 595.
2. Nevo,E. (1978) *Theor. Pop. Biol.,* **13**, 121.
3. Watt,W.B., Cassin,R.C., and Swan,M.S. (1983) *Genetics,* **103**, 725.
4. Watt,W.B. (1983) *Genetics,* **103**, 691.
5. Krieg,F. and Guyomard,R. (1983) *C.R. Acad. Sc. Paris,* **296**, 1089.
6. Karakousis,Y. and Triantaphyllidis,C. (1990) *Heredity,* **64**, 297.
7. Berg,G. (1962) *Freshwater Fishes of the USSR and Adjacent Countries.* IPST, Jerusalem.
8. Brown,W.M. (1985) In *Molecular Evolutionary Genetics.* In MacIntyre,R.J. (ed.), Plenum Press, NY.
9. Dawid,I.B. and Blacker,A.W. (1972) *Devel. Biol.,* **29**, 152.
10. Hutchison,C.A., Newbold,J.E., Potter,S.S., and Edgell,M.H. (1974) *Nature,* **251**, 536.
11. Wilson,A.C. and 10 co-authors (1985) *Biol. J. Linn. Soc.,* **26**, 375.
12. Brown,W.M., George,M., and Wilson,A.C. (1979) *Proc. Nat. Acad. Sci. USA,* **76**, 1967.
13. Cann,R.L. and Wilson,A.C. (1983) *Genetics,* **104**, 699.
14. Hoelzel,A.R., Hancock,J., and Dover,G.A. (1991) *Mol. Biol. Evol.,* **8**, 475.
15. Chang,D. and Clayton,D.A. (1985) *Proc. Nat. Acad. Sci. USA,* **82**, 351.
16. Johnson,M.J., Wallace,D.C., Ferris,S.D., Rattazzi,M.C., and Cavalli-Sforza,L.L. (1983) *J. Mol. Evol.,* **19**, 255.
17. Meyer,A., Kocher,T.D., Basasibwaki,P., and Wilson,A.C. (1990) *Nature,* **347**, 550.
18. Cann,R.L., Stoneking,M., and Wilson,A.C. (1987) *Nature,* **325**, 31.
19. Kreitman,M. and Aguadé,M. (1986) *Proc. Nat. Acad. Sci. USA,* **83**, 3562.
20. Hudson,R.R., Kreitman,M., and Aguade,M. (1987) *Genetics,* **116**, 153.
21. Kreitman,M. (1983) *Nature,* **304**, 412.

22. Coen,E.S., Thoday,J.M., and Dover,G.A. (1982) *Nature,* **295**, 564.
23. Szostak,J.W. and Wu,R. (1980) *Nature,* **284**, 426.
24. Arnheim,N. (1983) In *Evolution of Genes and Proteins*. Nei,M. and Koehn,R.K. (ed.), Sinauer, Sunderland, MA.
25. Brown,S.D.M. and Dover,G.A. (1981) *J. Mol. Biol.,* **150**, 441.
26. Ohta,T. and Dover,G.A. (1983) *Proc. Nat. Acad. Sci. USA,* **80**, 4079.
27. Flavell,R.B., O'Dell,M., Sharp,P., Nevo,E., and Beiles,A. (1986) *Mol. Biol. Evol.,* **3**, 547.
28. Nevo,E. (1983) In *Proceeding of the Sixth International Wheat Symposium*. Sakamoto,S. (ed.), Kyoto University Press.
29. Dover,G.A. and Flavell,R.B. (1984) *Cell,* **38**, 623.
30. Jeffreys,A.J., Wilson,V., and Thein,S.L. (1985) *Nature,* **314**, 67.
31. Wong,Z., Wilson,V., Jeffreys,A.J., and Thein,S.L. (1986) *Nucleic Acids Res.,* **14**, 4605.
32. Jeffreys,A.J., Neumann,R., and Wilson,V. (1990) *Cell,* **60**, 473.
33. Davies,N.B. (1983) *Nature,* **302**, 334.
34. Burke,T., Davies,N.B., Bruford,M.W., and Hatchwell,B.J. (1989) *Nature,* **338**, 249.
35. Westneat,D. (1990) *Behav. Ecol. Sociobiol.,* **27**, 67.
36. Davies,N.B. and Houston,A.I. (1986) *J. Anim. Ecol.,* **55**, 139.
37. Gilbert,D.A., Lehman,N., O'Brien,S.J., and Wayne,R.K. (1990) *Nature,* **344**, 764.
38. Hoelzel,A.R. and Dover,G.A. (1991) *Heredity,* **66**, 191.

Glossary

Allele: one of a series of possible alternative forms of a given gene differing in DNA sequence and affecting the structure and/or function of a single product (RNA and/or protein).

Allopatric: where populations, species or taxa occupy separate geographic areas.

Alu family: a short, interspersed DNA sequence repeated about 500 000 times in the human genome, and characterized by containing a distinctive *Alu*I restriction site.

Allozyme: allelic forms of an enzyme that can be distinguished by electrophoresis.

Anadromous: having the habit of migrating from salt water to fresh water to breed.

Assortative mating: non-random mating during sexual reproduction involving a tendency for males of a particular kind to breed with females of a particular kind.

Autozygosity: homozygosity such that the two homologous genes are identical by descent.

Bottleneck: fluctuations in allelic frequencies when a large population passes through a contracted stage and then expands again with an altered genetic composition as a consequence of genetic drift.

Chromatids: the two daughter strands of a chromosome after replication, joined by a single centromere.

Chromosome: structure containing DNA and proteins in the cell nucleus.

Codon: a triplet of nucleotides in a gene that specify a given amino-acid in a protein.

Conspecific: said of organisms that are of the same species.

'Control' region: sequences of DNA usually near the beginning of a gene that regulate the transcription of the gene.

5′, 3′ controls: the beginning and end of a gene are called 5′ and 3′ relative to the direction of transcription. Control regions can be at either end.

Copy number: the number of copies of a given gene in a set of chromosomes, see **multigene family**.

Crossing-over: the exchange of genetic material between chromosomes due to chromosome 'breakage' and reunion.

Cryptic simplicity: regions of DNA in which the frequency of given short DNA motifs (e.g. ATAG) is higher than expected by chance and in which several motifs are scrambled one with another, see **slippage** and **pure simplicity**.

Deme: a local group of individuals that interbreed.

Demographics: the study of populations, especially growth rates and age structure.

DNA: deoxyribonucleic acid, the molecular basis of heredity.

cDNA: complementary DNA made by reverse transcription of messenger RNA.

rDNA: the genes for several classes of ribosomal RNA molecules that go into the construction of ribosomes, usually in long tandem arrays in the chromosomes.

mtDNA: circles of DNA in the mitochondrion.

DNA reannealing: double-stranded DNA separates into single strands when heated which reanneal back into double strands when temperature is lowered.

DNA turnover: continual gain and loss of regions of DNA due to a variety of mechanisms of rearrangement; see **gene conversion**, **unequal crossing-over**, **slippage**, **transposition**.

Discriminant function analysis: a statistical method of assigning observations to groups based on previous observations from each group.

Exon: part of genes carrying genetic code for polypeptides; see **intron**.

F-statistic: a test used to evaluate the probability that two samples are drawn from the same population.

Fingerprinting: separation of the DNA of an individual into defined fragments the lengths of which are determined by the spacing of given restriction of enzyme sites. Numbers and lengths of fragments form a unique 'DNA fingerprint' for each individual.

Fixation index: the frequency of homozygosity for a particular allele at a given locus.

Flanking controlling sequences: see **5′, 3′ controls**.

Founder effect: see **bottleneck**.

Frameshift mutation: insertion or deletion of a nucleotide base in an exon such that the genetic code is read in a different frame.

G-test: a log-linear version of the chi-squared statistical test for goodness of fit.

Gene conversion: the ability of one allele of a gene (or one member gene of a gene family) to alter the sequence of another allele (or another member gene) to its own sequence. Usually occurs during meiotic recombination. For example *Aa* can become *AA* or *aa*. If conversion is biased there is a preference for *A* or *a*.

Gene conversion domain: a stretch of DNA involved in gene conversion which can vary from a few to thousands of base pairs in length.

Genetic distance: a measure of the number of allelic substitutions per gene that have occurred during the separate evolution of two populations or species.

Genetic marker: a sequence of DNA, usually recognizable by a restriction enzyme, that is diagnostic for a given chromosome.

Genome: the sum total of all the DNA on a haploid set of chromosomes in the nucleus of an individual, including both coding and non-coding sequences.

Genomic library: a collection of artificially cloned fragments representative of an individual's genome.

Hardy – Weinberg equilibrium: the rule stating that gene frequencies remain constant from generation to generation in an infinitely large, randomly interbreeding population with no selection, migration, or mutation.

Haplotype: a set of alleles of closely-linked genes that tend to be inherited together, uniquely identifying a chromosome.

Heteroduplex: formed when one strand of one DNA duplex invades and displaces one strand of another DNA duplex during meiotic recombination, forming a mixed duplex without correct A – T and G – C pairing all along its length.

Heteroplasmy: individuals carrying more than one type of mitochondrial or chloroplast DNA.

Heterozygosity: the condition of having a pair of dissimilar alleles at a locus.

Homogenization: the process, arising as a consequence of DNA turnover, which ensures that most member genes of a multigene family are very similar in sequence; see **molecular drive**.

Homologous: chromosomes that pair during meiosis and contain the same linear arrangement of genes.

Homoplasmy: individuals carrying only one type of mitochondrial or chloroplast DNA.

Homozygosity: the condition of having identical alleles at a locus.

Hypervariability: extreme genetic variations between individuals in certain genomic sequences, see **fingerprinting**.

Hybridization stringencies: the fidelity with which single strands of DNA reanneal depends on the stringency of hybridization determined by temperature and ionic conditions.

Intergenic spacer (IGS): a region of DNA separating classes of ribosomal RNA genes in tandem arrays.

Interlocus variance: differences between genes in the number and frequency of alleles in a population.

Internal spacer: see **intergenic spacer**.

Intralocus variance: the frequency distribution of alleles of a gene in a population.

Intron: region of DNA which separates exons but which does not code for polypeptides.

Isozyme: alternative forms of a compound enzyme which is composed of polypeptides coded by different genes.

Kilobase: a region of DNA 1000 base pairs in length.

Linkage disequilibrium: non-random association of genes in the gametes of a population: the tendency of certain alleles of one locus to occur with certain alleles of another locus with frequencies greater than expected by chance.

Locus: region of DNA, usually a gene.

Minisatellite: tandem array of from 10 to 50 copies of a non-coding length of DNA. Arrays on different chromosomes are usually with different numbers of repeated copies, giving rise to unique individual DNA fingerprints.

Mobile elements: lengths of DNA that can move from one position to another in the genome.

Molecular drive: a process which spreads mutant genes through a multigene family (homogenization) and through a sexual population (fixation) as a consequence of a variety of non-Mendelian DNA turnover mechanisms in eukaryotic nuclear genomes.

Monomorphic loci: genes represented by a single fixed allele in a population.

Motif: a short defined sequence of DNA or polypeptide.

Multigene family: a collection of identical or near identical genes in the genome. The numbers of gene copies and their distribution amongst chromosomes varies widely between species depending on the gene family in question.

Neutral allele: a mutation in a gene that has little or no effect on the reproductive success of the individual carrying the allele.

Non-genic sequence: the bulk of sequences in nuclear genomes which do not code for polypeptides.

Non-Mendelian segregation: frequencies of genetic variants amongst the gametes of an individual which are not in accordance with predictions based on Mendel's laws of inheritance. All mechanisms of DNA turnover lead to patterns of non-Mendelian segregation.

Nonsense codon: a DNA triplet of bases that does not code for an amino acid but serves as a termination signal during protein translation.

Non-synonymous substitution: a nucleotide substitution, usually in the first or second position of a codon, that causes a replacement of an amino acid in a polypeptide chain.

Nucleotide: one of the units (A,T,G,C) from which DNA polymers are formed.

Overdominance: heterozygote advantage. The situation in which the heterozygote genotype (*Aa*) is more fit than either the homozygous recessive (*aa*) or homozygous dominant (*AA*) genotypes.

Phylopatry: when offspring share their home-range with their parents; non-dispersal.

Point mutation: a mutation involving a single nucleotide substitution.

Polygny: mating system in which there is one male and several females.

Polymorphism: two or more genetically distinct types in the same interbreeding population.

Polyphyletic group: a group of species that are classified together, some of which are descended from different ancestral groups.

Probe: a length of DNA or RNA radioactively labelled used to locate similar sequences in a heterogeneous collection of sequences.

Pure simplicity: regions of DNA in which a given short motif (e.g. ATAG) occurs in a tandem array without interruption; see **slippage** and **cryptic simplicity**.

Random mating: any male mating with any female without preference, see **assortative mating**.

Recombinant DNA techniques: techniques capable of locating, cloning, and genetically manipulating genes and other DNA sequences.

Recombination: the occurrence of progeny with gene combinations other than those found in their parents, caused by the independent assortment of chromosomes and gametes in a sexual species or by crossing-over between chromosomes.

Restriction sites: short motifs of DNA capable of being recognized by a restriction enzyme leading to the cutting of the DNA molecule into separate fragments. Each restriction enzyme has a unique cutting site.

RFLP (Restriction fragment length polymorphism): mutations at restriction sites which cause a restriction enzyme not to cut the DNA at that site (and mutations that generate new restriction sites) lead to DNA fragments of different lengths amongst individuals.

Sampling variance: variability due to the small size of the sample.

Satellite DNA: long tandem arrays of repeated sequences, usually in millions of copies, generally located at centromeres and telomeres of chromosomes. Generally thought to be generated by unequal crossing-over.

Sibship: genetic relationships between members of a familial pedigree (e.g. cousins, siblings, etc.).

Single-copy genes: genes for which only two alleles exist (one from each parent) in a diploid cell.

Slippage: a mechanism of DNA turnover by which gains-and-losses occur of short motifs (usually less than 10 nucleotides) in a DNA helix leading to pure and cryptic DNA simplicity (see above).

Subrepeat: a tandem array of repeats within a larger repeating unit, usually also in tandem array (e.g. subrepeats within the intergenic spacers of rDNA—see above).

Synonymous substitution: mutations in a codon, usually at the third position, which do not lead to a change in amino-acid at the polypeptide level.

Tandem array: head-to-tail arrangement of repetitive genes or non-coding DNA in the genome.

Tetrad: the four cellular products resulting from meiosis in a single cell.

Transposition: see **mobile elements**. Duplicative transposition occurs when a given DNA region replicates and the extra copy moves to another position in the genome. Non-duplicative transposition occurs when the DNA region moves from one position to another: no extra copies are involved.

Unequal crossing-over: crossing-over at a time when two chromatids or two chromosomes are not fully aligned leading to the gain of DNA on one chromatid (chromosome) with an equivalent loss in the other. A process of gain and loss which can generate multigene families and maintain their homogeneity; see **DNA turnover**, **molecular drive**.

Index